The Architecture of Reality

Thirteen Equations That Built the Modern World

Aaqhil Hussain

To Alia

Contents

Author's Note

I grew up in Kanchipuram, a small town in South India.

On some evenings, I would look up and notice three stars in a neat row, almost directly above me. I did not know their name then. Years later, I learned they were Orion's Belt.

Power cuts were common where I grew up. Most people hated them. I secretly loved them. When the lights went out, the sky changed. The few stars visible above the rooftops became thousands, and the darkness made the universe feel suddenly close.

One question stayed with me more than any other.

How far away are they?

I came from a religious home, and a warm one. When I asked about the stars, the answer was often God. For many people around me, that was where the question stopped. For me, it never could.

Not because I was trying to argue. I simply wanted a different kind of answer. I wanted to know how far away the stars were. How old the Earth was. How anyone, standing on this small rock in the dark, could possibly know such things at all.

And that is what changed everything: the realisation that we can. With enough patience, curiosity, and care, the universe does answer. Not through authority. Not through faith. Through evidence, measurement, and mathematics.

I eventually learned how far away Orion's Belt is. I learned how we know the age of ancient bones, planets, stars, and galaxies. And slowly, I found my way to the equations in this book.

Are these the thirteen most important equations ever written? That is my argument, not an official verdict. Another person might choose a different thirteen. But

I will make the case for each one, and more importantly, I will try to show you how each one can be understood, without pretending that mathematics is magic.

That is the spirit of this book.

It is the spirit of a boy looking up at three stars in a dark sky and refusing to stop at *because it is.*

Let me show you what I found.

The Architecture of Reality

The Quote You've Already Seen

You have seen this one before.

> *"Everybody is a genius. But if you judge a fish by its ability to climb a tree, it will live its whole life believing that it is stupid."* - Albert Einstein

It is warm, it is kind, and it carries the name of the most famous genius who ever lived. So it must be true.

There is only one problem. Einstein never said it. No one has ever found it in a single letter, speech, or interview. The line appeared long after he died, and his name was stitched on because it sounds like something he might have said. That was enough. A good sentence, a great name, and millions of us repeated it without once asking: did he really?

We did not check. Not because we are stupid, but because the name was big enough to switch the question off.

And maybe, somewhere along the way, someone judged you the same way they judged that fish. A teacher, a test, one bad year told you that you were "not a numbers person." You believed it. Most of us did. It was never true. It was the wrong test, set by someone who could not explain the thing simply.

The Architecture You Cannot See

Look up from this page for a moment.

The roof over your head is not falling on you. That is not luck. It stays up because someone worked out exactly how force travels through wood and steel and stone, wrote it down as an equation, and trusted the answer enough to build under it. Every beam above you is holding its weight because the numbers said it would.

It does not stop there. The plane that holds you up at forty thousand feet stays in the air because of equations about how air moves. The phone in your hand knows where you are standing, to within a few metres, because of equations about space, time, and gravity, running quietly inside satellites far above you. That same phone carries the face of someone on the other side of the planet because of equations about light and information. The power in your wall. The engine in your car. The bridge you crossed this morning. And we finally understand why the Sun burns at all, and does not simply go out. Equations. All of it, equations.

You did not vote for them. You cannot see them. You will never be sent a bill. And yet they hold up your entire world, every second of every day, and ask for nothing in return.

This is what the title of this book means. Architecture. Not as a metaphor. The way hidden steel holds up a tower you thought was made of glass, a surprisingly small number of equations hold up the whole of modern life. Pull them out, and everything you trust quietly falls down.

There are not millions of them. This book is about thirteen. Thirteen short lines that, between them, do most of the heavy lifting of the modern world.

What This Book Is

You might have noticed a contradiction.

I said a moment ago that people do not question things, that a famous name makes us put our hands down. And yet these thirteen equations were questioned without mercy. Doubted, attacked, torn apart for years, sometimes for centuries.

So which is it? Here is the answer, and it is the whole

point of the book. It depends entirely on what is placed in front of you.

A quote hides behind a name. It asks you to trust the person instead of the claim, and so most of us never check. That is how a sentence Einstein never wrote can live for fifty years on a coffee mug.

An equation works the other way. It can carry the most famous name in history, and it does not matter in the slightest. If it is wrong, anyone can prove it wrong, and the name will not save it. Reality gets the final vote.

The fake quote was believed because it was never tested. These thirteen equations are trusted because they were tested to destruction, and refused to break.

That is why you can trust them with your life. The people who found them were called mad, mocked by the experts of their day, or died long before anyone believed them. Not one was accepted because a clever person said so. Each one had to walk through fire. And every single time, it held.

And here is the part that should change how you read

the rest of this book. You do not have to trust any of it. Not Einstein's word, not mine. You can check it yourself, with your own hands. The fish was judged by someone else. An equation lets you be the judge.

I will begin with the oldest one. It is almost four thousand years old, and still exactly as true as the day someone first scratched it into wet clay. Simple enough that a child can check it with a ruler. Important enough to build a civilisation on. And it begins, the way these things so often do, with a simple right angle.

Chapter One

Pythagoras' Theorem

The Square Corner

$$a^2 + b^2 = c^2$$

Look at the corner of the room you are sitting in. Where the wall meets the floor, there is an angle. It is ninety degrees. A right angle.

Now look around. The window. The door. The screen in your hand. The page in front of you. The table, the brick, the box, the building outside. Almost every straight edge in the human world meets another one at a right angle.

This is not decoration. The right angle is the most useful shape we have ever found. Square corners stack.

They stand straight. They carry their weight evenly and do not lean. Build with true right angles and your wall can stand for a thousand years. Build with corners that are a little bit off, and the small error adds up, floor by floor, until the whole thing leans, cracks, and one day comes down. The right angle is the quiet promise that holds a building together.

Which leaves the first builder who ever lived with a hard problem. How do you make a perfect right angle? Not a close one. A perfect one. You cannot trust your eye. Your eye will lie to you by a degree or two, and a degree or two is the difference between a house and a pile of rubble. You have no protractor. You have no laser. You have a patch of ground, a length of rope, and the need to get it exactly right.

The answer humans found is one of the oldest pieces of cleverness we know of. And you can do it tonight, with a shoelace.

The Rope

Take a length of rope. Tie knots in it, evenly spaced, so the rope is divided into twelve equal parts. Join the ends into a loop. Now take three pegs, or three friends, and pull the loop tight into a triangle, with sides of three parts, four parts, and five parts.

Something quietly magical happens. The corner where the side of three meets the side of four is a perfect right angle. Every time.

You did not measure an angle. You did not own a single tool that knows what ninety degrees is. You counted, three, four, five, and the right angle appeared on its own, as if the numbers knew something about the world.

They do. This is a secret builders have used for four thousand years, and still use today. Walk onto a building site this week and you may see someone checking a corner by measuring three units along one edge, four along the other, and making sure the line between the two ends is exactly five. If it is five, the corner is square. The builder might not know whose equation it is. They do not need to. The rope does not care whether you understand it. It just works.

Why This Is Strange

Before we open the rope up and look inside, sit for a moment with how odd it is that it works at all.

A right angle is the angle of total independence. Think about walking north. You can walk north for a thousand kilometres, and your position east does not change by a single step. North and east ignore each other completely. They do not overlap. They do not lean on each other. They share nothing. That is what a right angle really is: two directions that have nothing to do with one another.

Now look at the triangle again. The two short sides are exactly that. Side a runs one way. Side b runs off at a right angle to it, in a direction that knows nothing about a. They are two completely independent journeys. And side c, the slanted one, is simply the shortcut that joins their two far ends.

So here is the puzzle. If a and b point in directions

that have nothing to do with each other, why should their two lengths reach across the triangle and secretly agree on the exact length of c? They have no reason to. And yet they do. Every single time. Nudge a a little, and c shifts to match, as if the two independent sides were quietly talking behind your back.

They are. We just have to work out what language they are speaking.

From Lengths to Squares

The first thing you might try is to connect the three lengths directly. It does not go well.

Take a right triangle whose two short sides are exactly 1 and 1. How long is the slanted side? It is the square root of two, $\sqrt{2} = 1.41421356\ldots$, and the decimals run on forever and never settle into a pattern. Lengths simply refuse to add up neatly around a corner. Chase them and you fall into a pit of endless, ugly numbers.

So we do something that sounds like a detour and turns out to be the whole secret. We stop measuring lengths. We measure areas instead.

On each side of the triangle, build a real square, a square whose edge is that side. This is all the little raised "2" in the equation ever meant. "Three squared" is not advanced mathematics, and it is nothing to be afraid of. It means a literal square tile, three units wide and three units tall. Count the small squares inside it: nine. That is three squared. A shape you can see, draw, and hold.

Now build all three squares and count the tiles.

The square on the side of length three holds nine tiles. The square on the side of length four holds sixteen. The square on the side of length five holds twenty-five.

And here is the moment: nine plus sixteen is twenty-five.

The two smaller squares, added together, are exactly the big one. Not close. Exactly. The square sitting on the long slanted side holds precisely as many tiles as the other two squares combined. Written in the shortest possible form, that whole sentence becomes:

$$a^2 + b^2 = c^2$$

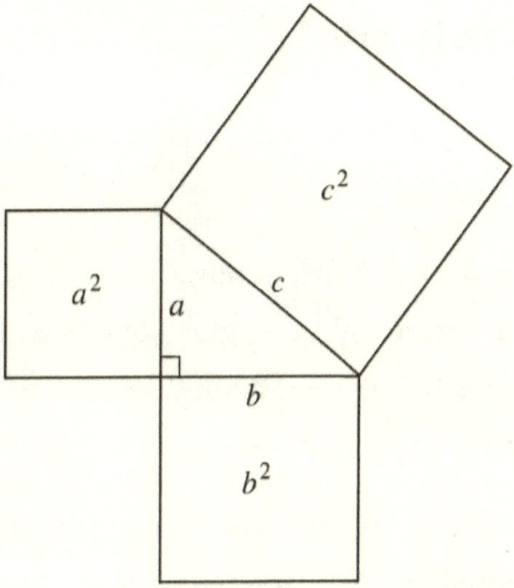

Figure 1: Figure 1.1 — The square on the longest side equals the two smaller squares combined.

Now look back at our puzzle. The two independent sides never did talk to each other directly. Their lengths had nothing to say to one another. But the squares they cast do. *Area* is the shared language the two directions speak. That is the secret handshake hiding inside every right angle, and inside the rope.

The lengths stay strangers. The squares always balance.

Is It Really True?

Now stop, and do the thing this whole book is about. Ask the question.

The rope works for three, four, five. Lovely. But does it work for *every* right triangle? For one with sides of seven and a bit? For one a hundred metres long? For one nobody has ever drawn?

Here is the trap, and it is a deep one. You could spend your entire life measuring right triangles. You could measure a thousand of them, ten thousand, and find that every single one obeys the rule. And you would still not have proved it. Because the very next triangle, the one you have not measured yet, might be the one that breaks it. Measuring tells you what *has* happened. It can never tell you what *must* happen. Ten thousand successful checks is still just a very long lucky streak, until you can show the thing could not have gone any other way.

So this is the real question. Not "does it work so far?" Anyone can see that it does. The question is harder: how could we ever be certain it works *always, everywhere,* for every right triangle that ever has been or ever will be drawn, including the countless ones we will never see?

That sounds impossible. It is not. And the way we close that gap is the single most powerful idea human beings have ever had. It is called a proof.

The Proof

A proof is a way of showing that something *cannot* be otherwise. Not "has not been." Cannot be. And the lovely thing about this particular proof is that you do not need a single line of algebra to follow it. You need four triangles and your own two eyes.

Here it is.

Cut four identical right triangles out of a piece of card. Any shape of right triangle will do, as long as all four are the same. Call the two short sides a and b, and the long slanted side c.

Now draw a large square, with each side as long as a and b put together.

Arrange the four triangles inside the big square the first way: tuck them in so that the empty space left over forms two smaller squares sitting next to each other. One is a square of side a. The other is a square of side b. The leftover space is $a^2 + b^2$.

Now, without changing the big square, and without trimming the triangles, simply slide them. Rearrange the same four triangles into a pinwheel, each one turned against the next, so their four slanted sides face inward. This time the empty space left over is a single tilted square in the middle. Its side is c. The leftover space is c^2.

Look at what just happened. The big square never changed size. The four triangles never changed size. You only slid them around. So the empty space they leave behind *must* be the same in both arrangements. It has to be. There is nowhere else for the area to go.

In the first arrangement, that empty space was a^2+b^2. In the second, it was c^2. They are the same space.

Therefore:

$$a^2 + b^2 = c^2$$

And now notice what you are feeling. You are not believing this because a clever Greek said so. You are not taking it on trust from a teacher, or from this book. You watched it. You could cut the card yourself and watch it again on your kitchen table. There is no triangle, anywhere, in any country, in any century, that can escape this, because the argument never leaned on the particular triangle you cut. It works for all of them at once.

That is what a proof does. It does not test a statement to destruction. It makes destruction impossible.

(For those who like the shorthand: the big square has area $(a+b)^2$*. That always equals the four triangles, which come to* $2ab$*, plus the empty space. First arrangement:* $(a+b)^2 = 2ab + a^2 + b^2$*. Second arrangement:* $(a+b)^2 = 2ab + c^2$*. Take the* $2ab$ *away from both sides, and* $a^2 + b^2 = c^2$ *falls straight out. Same proof, fewer pictures.)*

The Name on the Door

There is one last thing, and it brings us right back to where this book began.

We call this Pythagoras's theorem. Pythagoras was a Greek who lived around two and a half thousand years ago. Here is the awkward part. He almost certainly did not discover it.

The Babylonians were pressing these exact triangles into clay tablets more than a thousand years before Pythagoras was born. One of those tablets still survives. It is nearly four thousand years old, and it lists triangle after triangle that obey the rule, calmly, a full thousand years before any Greek wrote it down. The Indians knew it. The Chinese knew it. And Pythagoras himself? We know almost nothing reliable about the man. He led a secretive group who left little behind, and he may never have proved the theorem that now carries his name.

You have probably also heard that the ancient Egyptians used a knotted rope, the very rope we began with, to lay out the square corners of the pyramids.

It is a wonderful story. There is no real evidence for it. A historian guessed it in 1882, and the guess was repeated so many times it set hard, like plaster, into something that looks exactly like a fact.

Do you see the pattern? It is the same one from the very first page of this book. A line that famous lips never spoke. A story that never happened. Truths attract famous names and beautiful stories the way a magnet attracts iron filings, and a great many of the filings turn out to be wrong.

But here is what does not change, no matter which name we fix to it or which story we wrap around it. The triangle. Three, four, five still makes a right angle. The square on the long side is still exactly the two smaller squares combined. A Babylonian scribe, an Indian priest, a Chinese scholar, and a Greek philosopher all walked into the very same truth, separately, in different centuries and different languages, because it was never anyone's opinion to own. It was simply true, sitting there, waiting, the whole time.

Strip away every name. Strip away every story. The

theorem stands exactly where it always stood, completely unbothered. That is the difference between a fact and a story about a fact. The story needs a famous mouth to live in. The fact needs nothing from anyone. Reality cast its vote in Babylon, and in India, and in China, and in Greece, and it voted the same way every single time.

So Check It

Do not take Pythagoras's word for it. He may not have earned it. Do not take mine.

Find a piece of string and tie the knots. Or cut four triangles out of an old cereal box. The oldest equation in this book is also the one you can hold in your hands before you go to sleep tonight.

Make the corner.

It will be square. It always was.

But notice something about everything we have just done. It is all about a world that is holding perfectly still.

A right angle does not move. A square corner does not move. The three sides sit exactly where they are and never change. Pythagoras handed us the mathematics of things at rest. Of shape. Of a world frozen long enough to be measured.

And the world is almost never frozen. Right now the Earth is racing around the Sun at thirty kilometres every second. The air in this room is drifting. You shifted in your seat as you read that. Nothing in the universe truly holds still.

So the next question humans had to ask was not "what shape is it?" but "why does anything move the way it does?" Why does a loaded cart need a harder push than an empty one? Why does a thing, once moving, not simply stop on its own? Why does the world *go*, instead of just sitting there being a shape?

That question was harder than the right angle, and it waited far longer for its answer. When the answer finally came, from a difficult and brilliant man hiding from a plague, it folded into three of the most famous symbols ever written. You have seen them a hundred times. Almost no one has ever shown you what they truly mean.

$$F = ma$$

Chapter Two

Newton's Second Law

The Stubbornness of Things

$$F = ma$$

Imagine a shopping trolley.

It is empty, and you give it a push. It leaps forward, almost too eagerly, and you have to trot to keep up. Now picture the same trolley loaded to the brim, heavy with everything a week of groceries can hold. You push it with exactly the same effort as before. This time it barely stirs. It rolls forward slowly, grudgingly, as if it resents being moved at all.

You have felt this a thousand times and never thought about it. But you were touching one of the deepest laws in the universe with your bare hands. The same push, two different results, and the only thing that changed was how much stuff was in the trolley. The heavier it was, the more it resisted you. The more it resisted, the less it moved.

That feeling, that exact, ordinary, supermarket feeling, is the entire law of this chapter. Everything else is just learning to say it precisely.

$$F = ma$$

Three letters. You have seen them on a blackboard, in a film, on a t-shirt. Let us find out what they were trying to tell you the whole time.

The Two-Thousand-Year Mistake

Before we can read the law, we have to clear away a wrong idea, because it is probably sitting in your head right now, and it sat in the head of every human being for two thousand years.

Here is the wrong idea. *To keep something moving, you*

have to keep pushing it.

It feels completely obvious. Push a book across a table and it slides. Stop pushing, and it stops. Pedal a bicycle and it goes. Stop pedalling, and it slows and dies. The lesson of everyday life could not be clearer: motion needs a constant push to stay alive. Stop the push, and motion fades away.

The greatest mind of the ancient world, Aristotle, wrote this down as the truth. And because he was Aristotle, the world believed him for two thousand years. Two thousand years. Whole empires rose and fell while every educated person on Earth was confidently, completely wrong about why a ball rolls.

Because it is wrong. And the way we found out is one of the most important turns in this whole book.

The book stops, yes. But not because it ran out of push. It stops because the table is secretly pushing back. There is friction between the book and the wood, a gripping, dragging force you cannot see, fighting the motion the entire time. The bicycle slows because the air is shoving against you and the wheels are dragging on the road. The push was never keeping the motion alive. The push was only fighting off the *other* pushes that were trying to kill it.

The man who finally caught Aristotle out did not do it by arguing. He did it by checking. His name was Galileo, and in Italy in the early 1600s he did something almost no one had bothered to do in two thousand years of believing Aristotle. He went and tested it.

He rolled balls down smooth ramps, over and over, timing them, watching closely. And he noticed something teasing. A ball rolling down one ramp and up another always struggled back up to very nearly the same height it had started from, no matter how he angled the second ramp. Make the second ramp gentler, and the ball simply rolled further along it to reach that same height.

So Galileo asked a daring question. What if the second ramp were not a ramp at all? What if it were perfectly flat? The ball, forever trying to climb back to its old height and never finding any slope to climb, would just roll on. And on. And on. On a perfectly smooth, level surface, with nothing to slow it, the ball would never stop at all.

That was the death of Aristotle's idea, and it came from rolling balls down a plank. Galileo could not actually build a frictionless surface, no one ever has, but through the experiment he could *see* what would happen if he could. Motion does not need feeding. Left completely alone, it lasts forever.

We call this inertia, and it is so important that when Newton came along he made it his very first law, before the famous one we are here to meet. A thing at rest stays at rest. A thing in motion stays in motion, in a straight line, at a steady speed, forever, until a force says otherwise.

So what is a force actually *for*? Read the next sentence slowly, because everything turns on it.

A force is not what makes things move. A force is what makes moving things *change*.

What the Three Letters Mean

Now the law will read itself. Let us take the three letters one at a time, and you will find you already understand all of them.

Start with the last one, **a**. It stands for *acceleration,* and the only hard thing about acceleration is the name. It does not mean speed. It means *change* in motion. Speeding up is acceleration. Slowing down is acceleration. Turning a corner, even at a steady speed, is acceleration, because your direction is changing. Anything that alters how you are moving counts. And the opposite is just as important: if nothing is changing, if you are sitting still or gliding along at a constant speed in a straight line, your acceleration is zero. The law is not about motion. It is about the *change* in motion. Keep that in your pocket.

Next, **m**, for *mass.* Everybody thinks mass means weight, or how big something is. Closer to the truth is this: mass is how much a thing *resists being changed.* It is the trolley's stubbornness. A loaded trolley has more mass than an empty one, and that is exactly why it fought your push harder. A bowling ball has more mass than a balloon, which is why you can swat a balloon across a room with a finger but you cannot do the same to a bowling ball, no matter how annoyed you are at it. Mass is reluctance. It is how strongly a thing insists on carrying on exactly as it was.

And **F**, for *force*. That one you already know in your body. It is a push or a pull. Your hand on the trolley. Gravity on a falling apple. The engine driving a car. A force is anything that *tries* to change how something is moving.

Now put them together and the equation simply speaks:

$$F = ma$$

The force you apply equals the stubbornness of the thing, multiplied by the change you manage to produce in it. Push harder, and you get more change. But for the very same push, a more stubborn, heavier thing changes less. If you would rather see the result of the push instead of the cause, just shuffle the same equation around:

$$a = \frac{F}{m}$$

The change you get is the push, divided by the stubbornness. A big push on a light thing: enormous change. The same push on a heavy thing: barely a flicker. That is the loaded trolley and the empty one, written as mathematics. Your arms knew this equation long before your eyes ever read it.

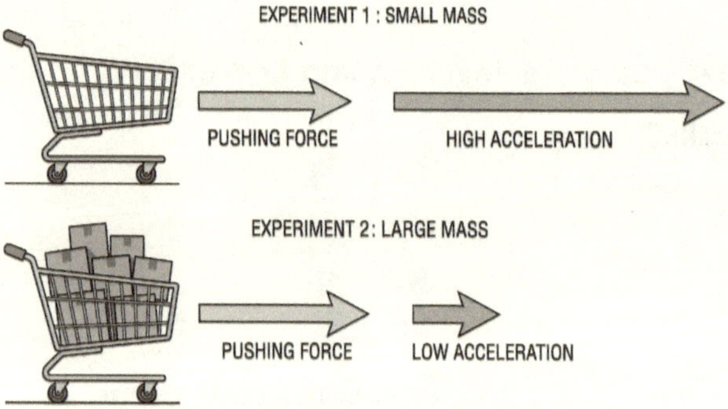

Figure 2: Figure 2.1 — Force, mass, and acceleration: the same push produces less change in a heavier object.

Mass Is Not Weight

One quick thing to settle, because almost everyone muddles it, and it will matter more than you would think.

I said mass is stubbornness: how hard a thing resists having its motion changed. Notice that this has nothing to do with gravity at all. Weight is the one that involves gravity. Weight is simply how hard gravity happens to be pulling on a thing at this moment.

Take a bowling ball to the Moon, where gravity is far weaker, and it *weighs* about a sixth of what it does here. It is easier to lift. But it is exactly as stubborn to shove sideways as it ever was on Earth, because its mass, the actual quantity of stuff inside it, has not changed by a hair. An astronaut floating in orbit weighs nothing at all, and yet a heavy toolbox up there is still horrible to get moving and just as horrible to stop. No weight, full stubbornness.

So mass and weight are two different things that happen to march in step here on the ground, which is the only reason we ever confused them. Mass is how much a thing resists being moved. Weight is how hard gravity pulls it down. Hold on to that difference, because the second half of it, gravity, is the road that leads out of this chapter and into the next.

The Thing That Is Never Lost

There is a hidden quantity living inside all of this, and it is worth meeting, partly because it is one of the most powerful ideas in all of physics, and partly because it is what Newton himself actually had in his head.

Take a moving thing, and multiply how much stuff it has by how fast it is going. Mass times velocity. Physicists call the result *momentum,* and the simplest way to think of it is as the total amount of motion a thing is carrying.

$$p = mv$$

A lorry crawling through a car park can carry more momentum than a rifle bullet, because although the bullet is blindingly fast, the lorry is so enormously heavier that it wins on the total. Momentum is the heft of motion: how much, moving how fast, all rolled into one number.

Now here is the remarkable thing about it. In any closed situation, momentum can be passed around, but it can never be created out of nothing, and it can never simply vanish. The total amount of motion before an event always equals the total amount after it. Always. The books must balance.

You can feel this for yourself, and it is genuinely strange the first time you notice it. Stand on a skateboard, perfectly still, holding something heavy, and throw it hard away from you. You will roll backwards. Where did that backward motion come from? A second ago you had none. It did not appear from nowhere. By giving the heavy object a parcel of forward momentum, you forced an exactly equal parcel of backward momentum to appear in yourself, so the total stayed at zero, exactly as it began. Step off a small boat onto a jetty and the boat scoots out from under your foot for the very same reason. Let go of an untied balloon and it darts crazily around the room, flinging air out of its neck one way and hurling itself the other.

This is the deep current flowing under $F = ma$. A force, it turns out, is really just momentum being handed from one thing to another.

Why Nothing Is Ever Lost

I keep saying momentum is never created or destroyed, only passed around. But this is a book about showing the working, so let us actually prove it, in three short steps, using nothing we have not already met.

First, look at the law of this chapter from a slightly different angle. Force is mass times acceleration, and acceleration is just how quickly velocity changes. So force is really how quickly *momentum* changes. Push on something, and the amount of momentum you pour into it is simply the force you apply multiplied by how long you apply it. Push for twice as long, deliver twice the momentum.

Second, recall that forces always come in equal and opposite pairs. When two things push on each other, the pushes are equal in size and opposite in direction. And they last for exactly the same length of time, because the shove is shared; it begins and ends for both at the very same instants.

Now put the two facts together and watch the result fall out on its own. Equal and opposite forces, acting for the identical length of time, must deliver equal and opposite amounts of momentum. So whatever momentum the one thing gains, the other loses exactly as much, in the opposite direction. Add the two together, and they cancel to nothing. The total momentum afterwards is precisely the total beforehand. Conservation of momentum was never a separate law we had to take on trust. It drops straight out of $F = ma$ and the equal-and-opposite rule, in three lines.

And you can put a number on it. Say you weigh 60 kilograms, you are standing still on a skateboard, and you hurl a 2 kilogram ball away from you at 10 metres per second. The ball carries off a momentum of its mass times its speed, $2 \times 10 = 20$. The books must balance from zero, so you must roll backwards carrying that very same 20 the other way. Your speed is that shared-out momentum divided by your own mass:

$$v = \frac{2 \times 10}{60} \approx 0.33 \text{ metres per second.}$$

A third of a metre per second, backwards, every single time. You can work it out before you throw, and on a slick enough floor, with a heavy enough object, you can go and feel the prediction come true under your own feet. The equation does not merely describe the recoil after it happens. It hands you its exact size in advance.

And this single idea, that motion is only ever handed on and never lost, is the secret behind one of the most impressive machines human beings have ever built.

Why a Rocket Works in Empty Space

Ask most people how a rocket flies, and they will tell you it pushes against the ground, or shoves against the air. It is a perfectly sensible guess. It is also completely wrong. And the proof is delightful: rockets work *best* exactly where there is no ground and no air at all, out in the empty vacuum of space.

So what on earth does a rocket push against?

Itself. Or, to be exact, its own fuel.

Newton noticed that forces never come alone. They always arrive in pairs, pointing in opposite directions. Push on a wall, and the wall pushes back on your hand with exactly equal strength; that is why your hand stops dead, and why, if the wall happened to be on wheels, it would roll away from you. Sit in a chair, and you press down on it while it presses up on you, holding you exactly in place. This pairing is so utterly reliable that Newton made it yet another of his laws: for every action, there is an equal and opposite reaction. Every push pushes back.

Now look at a rocket. A rocket is, at heart, a machine for throwing stuff backwards as violently and as fast as it possibly can. It burns its fuel and flings the searing exhaust gas out of its tail at tremendous speed.

That is the action: the rocket hurling its exhaust downward and backward. And the reaction, equal and opposite, is the exhaust hurling the rocket upward and forward.

The rocket never needed the ground. It never needed the air. It brought its own thing to push against, and carries it on board. It throws the fuel one way; the fuel throws the rocket the other; and out in the silent dark, with absolutely nothing else to touch, the rocket climbs.

It is precisely the skateboard and the heavy ball from a moment ago, scaled up until it can carry living human beings to another world. You throw something backward, you go forward. A rocket simply throws a colossal amount of something, extraordinarily fast, for a very long time. The crews of the Apollo missions rode all the way to the Moon on nothing grander than this: throw your fuel down hard enough, for long enough, and the books must balance by carrying you up.

Everywhere a Thing Moves

Pythagoras gave us the rule for a world standing still. This one is the rule for the world in motion, and the world is almost always in motion, which means this single short line is doing a staggering amount of the work that keeps you alive.

Every car ever built was designed with it. An engineer who knows the mass of a car and the force its engine can produce can tell you, before the car is even made, exactly how quickly it will speed up. Reverse it, and the same law tells you how much force the brakes must apply to stop that mass in time before it reaches the wall. Get the sum wrong, and people die. Get it right, and they arrive home.

It is the reason you survive a car crash that would once have killed you. Think about what an airbag and a crumple zone actually do. In a crash, your body has to change its motion fast, from highway speed to a dead stop. That change is the acceleration, and the force on your body is your mass times that acceleration. You cannot change your mass, and you cannot avoid stopping. But the law hands you one lever: *time*. If you can make the stop take longer, even by a fraction of a second, the change is spread out, the acceleration is gentler, and the force on your body drops. A crumple zone folds slowly instead of the car striking like a hammer. An airbag catches you over a long, soft moment instead of a steering wheel catching you in a hard, instant one. Both are the same trick, played straight out of this equation: stretch the time, soften the force, save the person. Engineers worked that out with $F = ma$, and it is quietly saving lives on every road on Earth, right now.

There is barely a moving thing in your world that this equation does not govern. The lift that carries you up a building. The ball a child throws. The blood your heart pushes through you. Your own body rising from a chair. If it moves, and if its motion changes, it is obeying $F = ma$, whether or not anyone involved has ever heard the name Newton.

What Acceleration Feels Like

Here is something strange, and worth sitting with for a moment: you cannot actually feel speed. You can only feel a *change* in speed.

Right now you are hurtling around the Sun at thirty kilometres every second, and you feel nothing whatsoever, because that motion is steady. A passenger dozing in a smooth aircraft at nine hundred kilometres an hour feels exactly as still as they would in their own bed. Pure, steady motion is invisible to the body. It is only acceleration, only *change*, that you can feel at all.

And feel it you do, constantly. Every lurch, every press, every flutter in your stomach is your body registering an acceleration, which is to say a force, which is to say $F = ma$ acting on you personally. When a plane charges down the runway and shoves you back into your seat, that shove is the seat applying exactly the force needed to drag your body up to the aircraft's new speed. When a lift starts upward and your stomach seems to drop for a moment, that is the floor pushing up on you harder than gravity, accelerating you, while your insides lag a beat behind.

Push the acceleration high enough and it turns dangerous. Fighter pilots hauling their jets through hard turns can feel the force on their bodies swell to several times their own weight, draining the blood from their heads until their vision greys out and they black out completely. Speed is not what does it; a pilot in a long, straight, steady dash feels perfectly fine. It is the violent *change,* the turn, the acceleration, the a in the equation climbing until the force on the body becomes more than it can take. The reason we measure these dangers in "g-forces" is that they are nothing more exotic than acceleration, doing exactly, and only, what this one short equation says it must.

Is It Really True?

Here we have to be careful and honest, because this law is a different *kind* of true from the last one, and the difference matters.

Pythagoras gave us a *proof.* We showed, with four triangles, that the theorem could not possibly be otherwise, in any universe, for any right angle, forever. It is locked. It needs no experiment, because it is true by pure reason.

$F = ma$ is not like that. It is not a theorem you can prove with logic alone. It is a *law of nature*: a claim about how the real, physical world actually behaves. And there is only one way to test a claim about the real world. You have to go and check. You have to push real trolleys, drop real weights, fire real rockets, and see whether the world does what the equation says.

So that is what people did, for three centuries, relentlessly, in every laboratory and workshop and launch pad on the planet. And the world obeyed, every single time, to a staggering precision. Few statements in all of science have been tested as often, or have passed as cleanly, as this one. That is *why* you can trust your life to a crumple zone. Not because a clever Englishman said so, but because the claim was checked to destruction and never once broke in the world we live in.

And yet, here is the honest part, the part most schools

never tell you. $F = ma$ is *not the final word.* Take a thing and push it to nearly the speed of light, and the law quietly bends. Its mass seems to swell, and the simple equation no longer fits. Shrink down to the size of a single atom, and the law dissolves into something far stranger, where particles do not even have one definite motion to change at all. We will meet both of those wilder worlds later in this book.

This is not a flaw. This is how the deepest kind of truth works. $F = ma$ is breathtakingly, reliably true across the entire world you will ever touch, the world of trolleys and cars and rockets and planets. And we now know precisely where its edges are, the doorways beyond which a deeper law takes over. Knowing exactly where a law stops being true is not a weakness. It is the highest form of understanding it. A law that has been pushed until it finally cracks, and whose every crack has been carefully mapped, is a law you can trust completely, so long as you respect its borders.

That is what it means for a law of nature to survive. Not that it is true everywhere and forever, like a theorem. But that wherever it claims to hold, it holds, every time you check, without exception, right up to the edge it never claimed to cross.

The Name on the Door

We call this Newton's Second Law, after Isaac Newton, and the story of how he found it is one of the great legends of science.

In 1665, plague swept through England, and Cambridge, where the young Newton was a student, shut its doors and sent everyone home. Newton went back to the quiet family farm at Woolsthorpe and spent roughly eighteen months more or less alone. In that year and a half, with nothing to do and no one to distract him, he did some of the most important thinking any human has ever done. He laid the foundations of calculus, took apart the nature of light and colour, and began chasing the idea of gravity. People call it his *year of wonders.* Most scientists would trade their whole career for that single isolated year of his.

But now, the careful thing, because this is a book

about careful things, and it is going to surprise you.

Newton never wrote $F = ma$.

It is true. Look in his great book, the *Principia* of 1687, and you will not find that equation anywhere. Newton did not even use "acceleration" the way we do. He wrote the law in words, as a proportion, and he framed it around *momentum,* that very "quantity of motion" we met a few pages ago: the change in a body's momentum, he said, is proportional to the force pressed upon it. It is the same idea, dressed in older and more careful clothes than the clean line we memorise today.

The tidy equation $F = ma$, the one on every blackboard, was crystallised later, in the 1700s, after Newton was dead. And the mathematician who did the most to forge it into that exact, beautiful, modern form was a Swiss genius named Leonhard Euler, whom we will meet properly in two chapters' time, and who has a quiet habit of handing the world the very symbols it then forgets he gave it.

So once again, the name on the door and the hand that

built the room are not quite the same. The most famous equation bearing Newton's name was finished by someone else, decades after he died. And, just as with Pythagoras, it does not matter even slightly. The law is true no matter whose initials we stamp on it. Newton saw the truth and described it. Euler sharpened the words. The trolley in the supermarket obeys it either way, and always did, long before either man was born to argue about who gets the credit.

So Feel It

You do not need to take Newton's word for this. He did not even write it the way you learned it. You do not need to take mine.

You can feel this one in your own body, today, no equipment required.

Find a chair on wheels, or a trolley, or a friend on a skateboard. Give it a push, then load it up and push again with the same effort. Feel the stubbornness change. That is the m fighting back. Stand on something that rolls and throw something heavy, and feel yourself drift the other way; that is momentum refusing to be lost. Next time you are a passenger in a car, pay attention to your own body as it speeds up, brakes, and turns. The press into your seat, the lurch forward, the lean into a bend, that is your own mass being forced to change its motion, and the push you feel is $F = ma$ acting on you personally, in real time. You are not reading about the law. You are sitting inside it.

But notice what the law does *not* tell you. It tells you, with perfect precision, what a force *does* once it acts: how much change a given push will produce on a given mass. It says absolutely nothing about where the force comes from in the first place. It is a law about effects, and it stays completely silent about causes. $F = ma$ will faithfully calculate the fall of an apple, but it has not the faintest idea *why* the apple falls at all.

And the most important force in your life is one you have never seen, never touched, and cannot switch off. It is holding you in your chair this very second. It held the apple to the tree, and then pulled it down. It reaches across empty, silent space to keep the Moon swinging around the Earth and the Earth swinging around the Sun, and it has done so, without rest, for billions of years.

Newton found it in that same plague year, and the story, as you may have heard, begins with an apple. Some of that story is true. Some of it is exactly the kind of beautiful tale this book has taught you to be suspicious of. I will sort the one from the other next.

Chapter Three

Universal Gravitation

The Apple and the Moon

$$F = G\frac{m_1 m_2}{r^2}$$

You know the cartoon. A young Isaac Newton dozes beneath a tree. An apple drops, bonks him squarely on the head, and in a flash of light he leaps up shouting that he has discovered gravity.

Almost none of that happened.

There was no blow to the head; that detail was added by later writers who could not resist improving the tale. He was not at Cambridge under some famous campus tree; there was no such tree. The "eureka" moment, the sudden flash, is a storyteller's flourish.

Like the Egyptian rope and the line Einstein never wrote, the story grew in the retelling until the polished version travelled far better than the plain one ever could.

Here is the plain one, the version Newton himself told an old friend over tea, more than fifty years later. During the plague years, hiding at his mother's farm, he was sitting in the garden when he watched an apple fall from a tree. And it set him wondering. Not the obvious thing.. Everybody already knew apples fall. He wondered something stranger.

Why does the apple fall *straight down*? Why not sideways, or upward? It always heads for the centre of the Earth, as if the whole planet were quietly reaching up and pulling it home. And if the Earth can reach up and grab an apple at the height of a tree, then how high does that reach go? Ten metres? A hundred? A thousand?

As high as the Moon?

That question is the most important one anyone asked in the seventeenth century, and to feel why, you have to understand what people believed about the sky before he asked it.

The Wall Between Earth and Heaven

For two thousand years, everyone knew that the heavens and the Earth were two completely different kinds of place.

Down here was the earthly realm: messy, changeable, corruptible. Things fell, decayed, broke, died. Up there was the celestial realm: perfect, eternal, untouchable. The Sun, the Moon, the planets and stars moved in flawless circles forever, made of some finer heavenly stuff, obeying their own divine rules. There was a wall between the two worlds, and it was unthinkable that the same grubby laws governing a falling apple could have anything to do with the serene clockwork of the heavens.

Newton's question quietly proposed to tear that wall down.

What if there is no second realm? What if the very same pull that drags an apple to the ground also reaches out, across a quarter of a million miles of empty space, and holds the Moon? What if the heavens are not perfect and separate at all, but are running on the exact same rule as a falling fruit?

This was not a small tweak to the physics of the day. It was the boldest claim a human had ever made about the universe: that it is *all one place,* top to bottom, governed end to end by a single law you could write on the back of your hand. The same rule for the apple and the angels. That is why we call it *universal* gravitation. Not gravitation that happens to be quite widespread. Gravitation that holds, without exception, for every speck of matter that has ever existed, anywhere.

And the way Newton tested this enormous idea was breathtakingly simple. He did the sum.

The Moon Is Falling

Here is the thought that cracked it open, and it is so strange and so beautiful that it is worth slowing right down for.

The Moon is falling.

Right now, tonight, the Moon is falling toward the Earth, dropping toward us exactly the way that apple dropped from the tree. This sounds insane, because the Moon very obviously does not crash into us. But stay with it, because the resolution is one of the most gorgeous ideas in all of science.

Throw a stone horizontally. It travels forward, but as it goes, gravity pulls it down, and it curves to the ground a few metres away. Throw it harder, and it goes further before it lands. Harder still, further still. Now imagine throwing it so monstrously hard that, as it falls, the ground itself curves away beneath it. The Earth is a ball, after all. It is always curving downward and out of the way. If your stone falls toward the Earth at exactly the rate the Earth's surface curves away, then it never gets any closer to the ground. It falls, and falls, and falls, forever, in a great circle around the world, always dropping, never landing.

That is an orbit. Newton imagined firing a cannonball from a mountaintop, faster and faster, until it stopped coming down at all and simply fell around the planet in a permanent circle. There is even a famous sketch of it in his work. An orbit, he realised, is not some special heavenly motion. It is just *falling, while moving sideways fast enough to keep missing the ground.*

The Moon is that cannonball. It is forever falling toward the Earth, and forever moving sideways quickly enough that it keeps missing. It has been missing for four and a half billion years. The apple, which has no sideways speed, simply falls and hits. The Moon, racing sideways at a kilometre every second, falls just as truly, and never arrives.

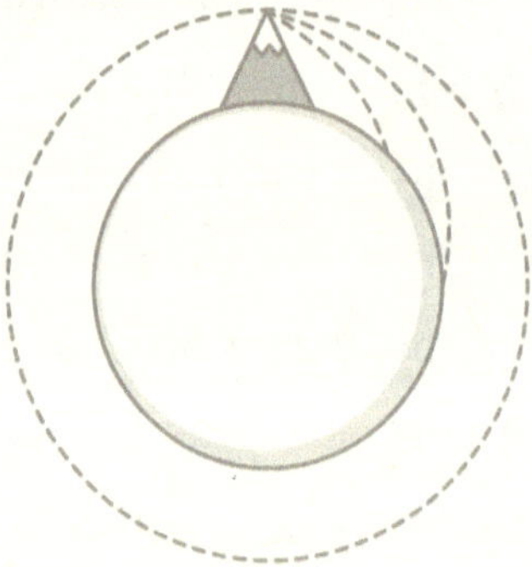

Figure 3: Figure 3.1 — Newton's cannon: at orbital speed, a projectile falls around the Earth and never lands.

Look back at the last chapter and you will see the two halves clicking together. The Moon *wants* to fly off in a straight line, because of inertia; left alone it would sail away into the dark forever. But gravity bends that straight line into a curve, tugging it inward, over and over, every instant. Caught between its sideways rush and the inward pull, it does the only thing it can. It goes in a circle. It is the rock on the string from a few pages ago, except the string is made of gravity and it reaches across empty space.

So Newton did the calculation. He worked out how fast the apple accelerates as it falls near the ground. Then he worked out how much the Moon "falls" away from a straight line in its orbit. And he asked whether one number could be turned into the other by a single, simple rule about distance. When the sums finally lined up, he had his answer, and the wall between Earth and heaven was gone forever. The same force. One law. From the orchard to the Moon.

The Law Itself

So what is the rule? Here it is, and like all the best ones, it is shorter than the sentence describing it:

$$F = G\frac{m_1 m_2}{r^2}$$

Do not flinch. I will take it apart and you will find every piece is something you already feel in your bones.

On the left, F, the force of gravity, the pull. On the right, the recipe for how strong that pull is between any two objects in the universe.

The two objects have masses, which we call m_1 and m_2. The law says you *multiply* them together. So the more massive either object is, the stronger the pull. A bigger planet pulls harder. A heavier you is pulled harder. This is why you are stuck to the Earth and not the Earth to you in any way you would notice: the Earth is so enormously more massive that its share of the pull dominates completely. But, and this is the universal part, *you are pulling on the Earth too,* with your own tiny mass, and you are pulling on the person sitting near you, and on the Moon, and on a distant star. Everything pulls on everything. Always. The pulls are mostly too faint to feel, but they are never, ever zero.

Then there is r, the distance between the two objects, and notice what the law does with it. It does not just divide by the distance. It divides by the distance *squared*. This is the famous inverse square law, and it has a beautiful, almost visual reason for being true.

Why the Square?

Imagine the pull of gravity spreading out from a planet like light spreading from a bulb, or sound from a speaker. It radiates outward in every direction at once, into an ever-growing sphere.

Now, the surface of a sphere grows with the square of its radius. Double the radius, and the surface is not twice as big, it is four times as big. Triple it, and the surface is nine times as big. So the same total amount of "pull," spreading out to cover that ballooning surface, gets thinner and thinner in exactly that proportion. At twice the distance, the same pull is spread over four times the area, so any one spot feels a quarter of the strength. At three times the distance, a ninth. That is what $\frac{1}{r^2}$ means. Move twice as far away, and gravity does not become half as strong. It becomes a *quarter* as strong. The pull fades fast, and it fades for a reason as simple as the geometry of a sphere. The universe is not dividing by distance arbitrarily. It is just spreading a fixed thing over a growing ball.

And finally, the quiet letter at the front: G. It is called the gravitational constant, and it is the number that sets the overall strength of gravity everywhere in the universe. It never changes. It is the same on Earth, on the far side of the galaxy, and in the first second after the Big Bang. We will come to its strange smallness in a moment. But first, the whole grand idea had to face a test, and it is one you can run yourself.

The Sum That Proved It

Newton did not simply declare that one law rules both the apple and the Moon. He checked it, with a calculation you can follow on a single sheet of paper, and it is the moment the whole grand idea either lived or died.

Begin with the apple. At the Earth's surface, things fall with an acceleration of about 9.8 metres per second squared. That is just how briskly Earth's gravity speeds things up down here at ground level.

Now the Moon. The Moon sits about sixty times farther from the centre of the Earth than that apple does. And the law insists the pull weakens with the *square* of the distance. So out at the Moon, the pull, and therefore the rate of falling, should be feebler by a factor of sixty squared, which is $60 \times 60 = 3600$. If the law is right, the Moon ought to be falling toward the Earth at

$$\frac{9.8}{3600} \approx 0.0027 \text{ metres per second squared,}$$

and not a whisker more or less. That is a hard, specific, riskable prediction.

So now go and measure the Moon, with no apples involved at all. We know how far away it is, and we know it takes about 27.3 days to circle the Earth once. From nothing but those two facts, you can work out how sharply its path must curve inward to hold that orbit, which is to say how fast it is falling away from the straight line it would otherwise fly off along. Do that arithmetic, and out comes a number:

0.0027 metres per second squared.

The same number. The fall of an apple in an English orchard, divided by sixty squared, lands exactly on the curve of the Moon a quarter of a million miles away. One law, measured against two things at opposite ends of that enormous gap, agreeing to the decimal place. This is the sum Newton did in his plague year, and when the two numbers matched, the ancient wall between the earthly and the heavenly was rubble.

You can do the first half this minute, on any calculator: nine point eight, divided by three thousand six hundred. That is your prediction for how fast the Moon falls. The night sky has been quietly confirming the second half for three and a half centuries, and it will confirm it for you tonight.

Gravity Is Pathetically Weak

We think of gravity as mighty. It holds planets, crushes stars, swallows light into black holes. So it comes as a shock to learn that gravity is, by an absurd margin, the *weakest* force in nature.

You can prove this to yourself with a fridge magnet. Stick a paperclip to it. That little magnet, costing a few cents, is holding the paperclip up against the gravitational pull of the *entire planet Earth*, six thousand billion billion tonnes of it, and the magnet wins easily. The whole Earth pulls down on that paperclip, and a toy magnet overrules it without effort. Gravity is that feeble.

So how on Earth does something so weak end up running the universe?

Because it never gives up, and it never cancels out. The other forces are stronger, but they come in opposites that neutralise one another; pile up enough matter and the pushes and pulls largely cancel, leaving nothing felt at a distance. Gravity has no opposite. There is no negative mass, no anti-gravity lurking in ordinary matter. Every scrap of mass pulls, and only pulls, and all those tiny pulls add up, and up, and up, the same direction every time, never once cancelling. One apple's pull is nothing. But a whole planet's worth of apples, all pulling together, becomes a force that keeps the Moon on its leash. Gravity wins the universe not by being strong, but by being relentless, and by always, everywhere, pulling the same way.

Twice a Day, the Sea

Here is that same feeble force doing something colossal, on a timetable you can check against a tide chart.

The Moon, our falling cannonball, is pulling on the Earth. But it does not pull on every part of the Earth equally, and that small inequality is the entire story of the tides. The side of the Earth facing the Moon is a little closer to it, so it is tugged a little harder. The far side is further away, so it is tugged a little less. We just saw why: gravity weakens with distance, by that inverse square, so the near side always feels a stronger pull than the far side.

Now think about the water. On the side facing the Moon, the ocean is pulled toward the Moon more strongly than the solid Earth beneath it, so it bulges outward, toward the Moon. That much you might have guessed. But here is the part that surprises almost everyone, and that explains something you have half-noticed your whole life: on the *opposite* side of the Earth, the water bulges out as well, away from the Moon.

Why? Because over there, the Moon pulls the *Earth itself* away from the water more strongly than it pulls that far water. The solid planet gets dragged toward the Moon and slides out from under the far-side ocean, leaving that water behind in a second bulge, pointing straight away from the Moon. So the oceans pucker into two bulges at once, on exactly opposite sides of the planet, one reaching toward the Moon and one left behind it.

And now the Earth spins. As it turns through a day, any stretch of coastline is carried first through one bulge, then through the low water between, then through the *other* bulge, then the low again. Two bulges, one rotation, and so two high tides and two low tides, every single day, on every shore in the world. That is why the sea climbs the beach and falls back twice between one sunrise and the next. It is the inverse square law, written in water, on every coast on Earth.

The Sun joins in too, with its own smaller tides. When Sun and Moon line up, their bulges add into the especially high *spring* tides; when they sit at right angles, they partly cancel into the gentler *neap* tides. The whole breathing rhythm of every coastline on the planet is nothing but gravity, pulling a fraction harder here than there.

The Man Who Weighed the Earth

There is a lovely problem buried in that little letter G. To use Newton's law to weigh a planet, you first have to know the value of G, and for over a century after

Newton, nobody did. The law was perfect, but it had a blank in it.

Then, in 1798, an intensely shy English gentleman named Henry Cavendish, working almost alone, built one of the most delicate experiments anyone had ever attempted. He hung two small lead balls from a fine thread, and brought two much larger lead balls near them, and then watched, with almost unbelievable patience and care, for the threadful of small balls to twist by the tiniest fraction as the big balls gently, gravitationally, tugged them. From that minuscule twist, he measured the strength of gravity between lumps of metal in his own laboratory, and so, at last, pinned down G.

And the moment he had G, the law gave him something staggering in return. Knowing how hard the Earth pulls on things at its surface, and now knowing G, he could rearrange Newton's equation and solve for the one unknown left: the mass of the Earth itself. A quiet man in a private laboratory, twisting a thread with lead balls, calculated the weight of the entire planet beneath his feet. He got it very nearly right.

That is the power of a true law. Measure something tiny on a bench, and it hands you the mass of a world.

The Equation That Found a Planet

But the greatest thing this law ever did was not to measure something that was there. It was to find something that was not.

By the 1840s, astronomers had a problem. The planet Uranus was not where it was supposed to be. They had used Newton's law to predict its path around the Sun, and the planet stubbornly refused to follow it, drifting a little ahead here, lagging a little behind there. Something was wrong.

Now, here is the fork in the road, and it is the whole spirit of this book. They could have concluded that Newton's law was broken. Instead, two mathematicians, working separately, trusted the law completely, and asked a different question. What if the law is perfectly right, and there is simply *something else out there* pulling on Uranus, an unknown planet, unseen, whose gravity was nudging Uranus off course?

So they used the law in reverse. From the exact way Uranus was misbehaving, they calculated where this hidden planet would have to be to explain it. One of them, the Frenchman Le Verrier, sent his prediction to an astronomer with a good telescope. On the very first night of looking, in 1846, the astronomer pointed his telescope at the patch of sky the equation specified, and there, within a whisker of the predicted spot, sat a new planet, never before seen by human eyes. We call it Neptune.

Sit with that. A few symbols on paper, fed the wobble of one planet, pointed into the empty dark and said *there is a world right there,* and there was. Nobody had seen it. Nobody knew it existed. The mathematics knew first. That is what it means for an equation to be tested to destruction and refuse to break: you push it to predict something nobody has ever observed, and the universe turns out to have quietly agreed with it all along.

Surfing on Gravity

There is one more trick this law makes possible, and we lean on it every time we send a spacecraft to the outer planets, because we could never carry enough fuel to reach them any other way.

It is called a gravity assist, or a slingshot, and it sounds like cheating. A spacecraft falls toward a planet, whips around it, and comes out the far side travelling *faster* than it went in, having burned no fuel at all. Speed, apparently for free. Where does it come from?

It comes from the planet, and it is the momentum idea from two chapters ago, played on a colossal stage. A planet is not sitting still; it is racing around the Sun. When a tiny spacecraft swings close behind a moving planet, the planet's gravity drags it along, flinging it forward and handing it a sliver of the planet's own orbital motion. The books still balance, exactly as they must: the craft speeds up, and the planet slows down by precisely the amount needed to pay for it. But the planet is so unimaginably heavier that the speed it loses is far too small ever to measure. The craft gains kilometres every second. The planet loses less than the width of an atom's whisper. It is a flea stealing a scrap of speed from a charging elephant, and the elephant never feels it.

This is how the two Voyager probes, launched in the 1970s, toured the outer Solar System. They leapt from Jupiter to Saturn and onward, each giant planet whipping them faster, a chain of free kicks that eventually flung them clean out of the Solar System and into the space between the stars, where they still coast today, decades later. Every one of those slingshots was Newton's law and a borrowed scrap of momentum, navigated to the metre. We surf across the Solar System on the gravity of passing worlds.

The Thing Newton Could Not Explain

And yet, for all its triumph, this law hides a mystery so deep that Newton himself was disturbed by it, and was honest enough to say so.

Look at the equation again. The Sun pulls the Earth

across one hundred and fifty million kilometres of empty space. There is nothing in between. No rope, no rod, no medium, nothing touching. And yet, somehow, the pull is there, instantly, across all that nothing. *How?* How does one lump of matter reach across an empty void and tug on another, with nothing whatsoever to carry the pull?

Newton had no idea. And to his great credit, he refused to pretend otherwise. In a letter he called the very notion that gravity could act across empty space, with nothing to transmit it, so great an absurdity that he believed no thinking person could accept it. He had written the most successful equation in the history of science, an equation that found a planet, and he flatly admitted he did not know what gravity actually *was* or how it reached across the dark.

When pressed for a mechanism, he gave one of the most famous and honest lines in science. *I frame no hypotheses.* In other words: I can tell you, with perfect precision, exactly what gravity *does*. I have not the faintest idea what it *is*. And I will not insult you by making something up.

That gap, that beautiful, honest, unfilled gap, sat open for over two hundred years. The answer, when it finally came, would be even stranger than the question, and it would come from a clerk in a patent office who decided that gravity is not a force reaching across space at all. But that is a later chapter.

Is It Really True?

So how true is it?

For the entire world you will ever live in, staggeringly, reliably, life-and-death true. Every satellite over your head, every GPS fix, every weather forecast, every spacecraft we have ever flown, every prediction of an eclipse centuries in advance, all of it runs on this one equation, and it delivers, every single time. We flew human beings to the Moon and back on Newton's gravity. It is as trustworthy as anything humans have ever written down.

And you can watch its deepest promise confirmed with your own eyes. The law, combined with the last chapter's, makes a peculiar prediction: that *all* objects, heavy or light, should fall at exactly the same rate, because although a heavier thing is pulled harder, it is also, by precisely the same factor, more stubborn to move. The two effects cancel perfectly. A hammer and a feather should fall together. On Earth the air ruins the test, snatching at the feather. So in 1971, an astronaut standing on the airless surface of the Moon held out a hammer and a falcon feather and let them go. They fell side by side and struck the lunar dust at the very same instant. It was filmed. You can watch it. A feather and a hammer, falling as one, on another world, exactly as the equations promised three centuries before.

But, as always, the honest part. Newton's law is not the final word. There is one place in the Solar System where it very slightly fails: the orbit of Mercury, closest to the Sun, drifts by a tiny amount that Newton's law cannot quite account for. For a long time people assumed another hidden planet, another Neptune, was tugging at it. This time there was no planet. The law itself was very gently wrong, in the fierce gravity near the Sun. The fix would not be a new planet but a whole new picture of what gravity *is*, the picture

Newton could not supply, in which gravity is not a pull across empty space but the bending of space itself. We will get there.

This is the pattern of every great law in this book. Breathtakingly true across the world we inhabit. Tested to destruction and triumphant. And carrying, at its faint edges, the doorway to something deeper. Knowing exactly where Newton's gravity stops being perfect is not a mark against it. It is the most complete way to understand it.

The Name on the Door

We call it Newton's law, and unlike F = ma, this one he genuinely did write and prove, in his great book of 1687. But even here, the name on the door is not the whole story.

Newton stood on the shoulders of others, and said so. Two of those shoulders belonged to a pair of obsessives who lived a lifetime before him.

The first was Tycho Brahe, a Danish nobleman so devoted to the night sky that he built an observatory on his own island and spent decades recording the exact positions of the planets, night after night, by eye, before the telescope had even been invented. He was a colossal character, who had lost part of his nose in a youthful duel and wore a metal replacement for the rest of his life. But his measurements were the finest the world had ever seen, accurate to a hair, the best naked-eye data anyone had ever gathered.

When Tycho died, that priceless mountain of numbers passed to his assistant, a poor, brilliant, half-mystical German named Johannes Kepler. And Kepler did something heroic with it. He took the motion of Mars, the most stubborn planet of all, and tried for years to force it onto a circle, because everyone *knew* the heavens moved in perfect circles. It would not fit. Tycho's flawless data simply refused. So Kepler did the unthinkable: he gave up the circle. The planets, he found, move in ellipses, slightly squashed loops, speeding up when near the Sun and slowing when far, all in precise mathematical step. He had found the exact shape of the heavens.

But Kepler had only found *that* the planets move so.

He could not say *why*. That was the gift he handed across the decades to Newton, who took Kepler's hard-won laws and showed they fall straight out of a single rule about an inverse square. Without Tycho's patient eyes and Kepler's stubbornness, there would have been nothing for Newton's law to explain. The apple gave Newton the question. Kepler and Tycho gave him the answer to test it against.

And there was a bitter rival, Robert Hooke, a brilliant and prickly man who claimed, loudly and to the end of his life, that he had suggested the inverse square idea to Newton first. The two despised each other. When Newton wrote his famous line about standing on the shoulders of giants to see further, some have long suspected it was less humility than a quiet dig, since Hooke was notably small and stooped, and "giants" was exactly what he was not.

So even the cleanest credit in this chapter comes tangled with an old wound, a feud, a half-remembered apple, and a story polished smooth by admirers. The usual human mess clings to the law like dust. And the law does not care. Uranus wobbled, Neptune was found, the hammer and the feather fell together on the Moon, whatever anyone felt about anyone else. The truth sat there the entire time, indifferent to whose name we carved above the door.

So Look Up

You cannot weigh the Earth on your kitchen table. But you can do something nearly as good. You can go outside tonight and look at the Moon, and now you will see it differently.

It is not hanging there. It is falling. This very second it is dropping toward you, exactly as a thrown stone drops, and the only reason it does not land on your head is that it is also racing sideways fast enough to keep missing the Earth, over and over, forever. The same pull that would bring an apple down on your head is what holds that enormous bright stone in the sky. One law. The orchard and the heavens. You are looking at the proof, and it is the most ordinary thing in the world: something falling.

You do not need to take Newton's word for any of it.

He himself could not tell you what gravity *is*. Drop two coins of similar size together and watch them land as one, the way the hammer and feather did on the Moon. Feel your own weight pressing into your chair, the whole planet pulling on you, and you, faintly, pulling back on the whole planet. You are inside the equation, right now, as you always have been.

But leave this chapter with one last thing turning over in your mind. A handful of symbols, scratched on paper, were once aimed at a blank patch of sky and announced that an unseen world was hiding there. And the world was there. Think about how astonishing that is. Why should the universe pay the slightest attention to marks we make on a page? Why should reality bend to our mathematics at all?

That question is the deepest one in this entire book, and we have been circling it since the first page. The next chapter is the strangest clue we have ever found to the answer, where five numbers, discovered by five strangers who never met, fall into a single line so perfect that it suggests mathematics was never something we invented at all. We have been measuring the world. Now we are going to find out, just for a moment, that the world might have been built out of the very thing we were measuring it with.

Chapter Four

Euler's Identity

The Five Strangers

$$e^{i\pi} + 1 = 0$$

The last equation built things. Roofs, corners, walls, the square edge of every room you have ever stood in. This one builds nothing.

You will never use it to put up a shelf or land a plane. No engineer reaches for it in a crisis. You could delete it from the universe tomorrow and not a single bridge would fall. And yet, if you walk into a room full of mathematicians and ask them to name the most beautiful thing in their entire subject, a great many of them will write down this one short line, and then go quiet.

A poll of physicists once asked them to vote for the greatest equation ever written. This line tied for first place. People who spend their lives swimming in mathematics have compared it to a Shakespearean sonnet, to a perfect piece of music, to a painting that catches something about being alive that ordinary words cannot reach. One mathematician called it, simply, the truest beauty he knew.

That is a strange thing to say about five symbols and an equals sign. So this is going to be a different kind of chapter from the last one. Pythagoras gave us a truth that holds up the physical world. This equation gives us something stranger and, in its way, larger: a truth that holds up *mathematics itself*. It is the moment a person discovers that the subject they were taught in school as a pile of separate, unrelated, slightly boring rules is secretly one single thing. All joined up underneath. Connected in a way nobody had any right to expect, and nobody can fully explain even now.

To feel why grown adults get emotional over this, we first have to meet the cast. There are five of them. And the most important thing about them, the thing to hold on to through this entire chapter, is how little they have to do with one another.

Meet the Five

Five numbers live in that equation. Each one was discovered by different people, in a different century, on a different part of the planet, for a completely different reason. None of them was looking for the others. None of them had any reason to be in the same room. Let us introduce them one at a time.

Zero

Start with nothing.

It seems too obvious to be an invention. But zero, the number, was one of the hardest ideas humanity ever had to fight for. For most of history, people counted things, and you cannot hold up zero apples. The mighty Romans, who built roads and aqueducts that still stand, never had a symbol for zero at all. Try doing arithmetic with Roman numerals and you will feel the hole where it should be.

The breakthrough came from India. More than a thousand years ago, Indian mathematicians did something radical: they treated nothing as a number in its own right, a thing you could add, subtract, and calculate with. Around the year 628, a mathematician named Brahmagupta wrote down the rules for how zero behaves. It sounds small. It changed everything. Zero is what lets us write 10, and 100, and 1,000,000. Every large number you have ever read sits on top of an idea that took civilisation thousands of years to be brave enough to accept.

So zero is not boring. Zero is the result. The empty hand. The score before the game begins. The place everything returns to. Hold that thought, because this whole equation is going to end, exactly, on zero.

One

Then there is one.

One is the first something. The single unit. The beginning of counting, and the thing every other whole number is built from. Two is one and one. Three is one and one and one. One is the seed of all of arithmetic, so simple that we almost never stop to look at it.

Zero and one. Nothing and something. Already we have the two most basic ideas a mind can have. Keep them in view. They are both going to walk into the final line.

Pi

Now π, which you say like the word "pie."

Take any circle at all, a coin, a wheel, the Moon. Measure the distance all the way around it. Then measure the distance straight across it. Divide the first by the second. You always get the same number, no matter how big or small the circle is: 3.14159, and then the decimals keep going, and going, and never stop, and never settle into a repeating pattern. They are still being calculated by computers today, to trillions of digits, and they never end.

People have been chasing π for thousands of years. The ancient Greek Archimedes, more than two thousand years ago, trapped it between two many-sided shapes and squeezed until he had it pinned to within a hair. π comes from the simplest, roundest shape there is. And yet, as we are about to see, it shows up in places where there is no circle anywhere in sight. That is the first clue that something deeper is going on.

π is the number from circles. The constant of round things.

e

The fourth stranger is e, which is 2.71828, and like π its decimals run on forever without repeating.

e is the number from *growth*. We are going to build it ourselves, from nothing but a bank account and a bit of greed, in the next section. For now, just know this: anywhere something grows by feeding on what it has already grown, e is hiding in the mathematics. Money earning interest. A population breeding. A rumour spreading. A hot drink cooling down. e is the fingerprint of natural growth, and people first stumbled onto it while counting money.

i

And then there is the fifth one. The strange one. The one that, for nearly three hundred years, the cleverest people alive insisted was not a real number at all.

It is called i, and it was born out of an argument, an insult, and a very long misunderstanding. Its story is so good, and so perfectly on the theme of this book, that it gets its own section later. For now, all you need to know is that i was treated as a ghost, a fiction, an impossible number that mathematicians used while holding their noses, certain it could not really exist.

They were wrong. i is as real as 5 or minus 3. It was only ever the name that lied.

Look at Them Together

So here is our cast. Nothing. The first something. A number from measuring circles. A number from counting money. And an "impossible" number from solving equations that seemed to have no answer.

Five strangers. Born in five different cities, across five different millennia, found by people who never met, for reasons that had nothing to do with each other. A surveyor with a piece of string. A banker with a ledger. An Italian solving a puzzle in a duel. The plain, ancient ideas of nothing and one.

Now watch what happens when you put all five in a single line:

$$e^{i\pi} + 1 = 0$$

They balance. Perfectly. Exactly. The whole thing lands on zero with no remainder, nothing left over, nothing fudged.

Either that is the greatest coincidence in the history of the human mind, or these five strangers were family the entire time and simply had never been introduced. It is the second one. The rest of this chapter is us introducing them, and watching them recognise each other.

Building e From a Bank Account

Let us throw out the textbook definition of e and grow the number ourselves, out of pure greed.

Imagine you have one dollar. You find a gloriously generous bank that offers you 100 percent interest in a single year. Easy enough. At the end of the year your one dollar has doubled into two. Done.

But now you get clever. You go back to the bank and say: do not give me all the interest at the very end. Give me half of it halfway through the year, and let that sit and earn interest too. The bank shrugs and agrees. So at the six-month mark you collect 50 percent on your dollar, which leaves you holding one dollar and fifty cents. Then for the second half of the year, the full 50 percent interest is calculated on that *larger* amount. You finish the year with two dollars and twenty-five cents.

Look at that. You made an extra twenty-five cents, out of nowhere, simply by letting your interest start earning its own interest. That is the whole idea of compounding: growth that grows on its own growth.

So, naturally, you get greedier. Pay me four times a year, you say. Now the interest compounds four times, and you finish with about two dollars and forty-four cents. Better. Pay me every month, twelve times. You end with about two dollars and sixty-one cents. Every single day, 365 times? About two dollars and seventy-one cents.

You can feel where your mind wants this to go. Every time you slice the interest finer, you end up with more. So surely, if you could compound every hour, every second, every instant, continuously and forever, your dollar should explode off toward infinity.

It does not. And this is the beautiful, surprising part. The faster you compound, the smaller each new gain becomes. The numbers crawl upward and then they stop, pressing against an invisible ceiling, as if the universe simply refuses to let pure growth run away with itself. That ceiling is a specific, exact number:

$$2.71828182845\ldots$$

That is e. It is the precise result of growth compounding as fast as it possibly can, every instant, with no pause ever. It is not a decimal somebody invented. It is a wall built into the structure of growth itself.

And here is what makes e one of the five strangers and not just a banking trick. The exact same number governs a lump of uranium decaying over thousands of years. It governs how fast your coffee cools to room temperature. It governs how a population of rabbits multiplies, how a disease spreads through a city, how the charge drains out of a battery. Anywhere in the universe where the speed of change depends on how much is already there, e appears, uninvited, every time. People found it in a bank. It turned out to be everywhere.

That is stranger number four, built from a single dollar.

The Long Insult: The Story of i

Now the ghost. And to understand why everyone was so afraid of it, we have to go to Italy, five hundred years ago, into a world where mathematics was a blood sport.

In the early 1500s, mathematicians did not publish their best methods. They hoarded them like magicians guarding tricks, because they made their living from public duels. Two mathematicians would each hand the other a list of problems, meet in a town square, and whoever solved more in front of the crowd won money, students, and reputation. The great prize of the age was a method for solving the cubic, a certain kind of equation that had defeated everyone for centuries.

The method was found, lost, stolen, and fought over in one of the great soap operas of mathematical history, involving a stammering genius nicknamed Tartaglia ("the stammerer") and a brilliant, scandalous gambler-physician named Gerolamo Cardano. Cardano eventually published the solution, in a famous book called *Ars Magna* in 1545, breaking a promise of secrecy and earning Tartaglia's hatred for life.

But here is the part that matters for us. When Cardano used his shiny new method, it sometimes forced him to do something that looked insane. In the middle of a calculation, he would have to take the square root of a negative number.

This was supposed to be impossible. The square root of a number is the value that, multiplied by itself, gives you that number. Two times two is four, so the square root of four is two. But what is the square root of *minus* one? It cannot be a positive number, because a positive times a positive is positive. It cannot be a negative number, because a negative times a negative is also positive. There seemed to be no number anywhere that, multiplied by itself, gives a negative. The square root of minus one had no home on the number line at all.

Cardano stared at these things appearing in his work and did not know what to do with them. He wrote that they were as subtle as they were, in his own words, useless, and he backed away.

Then, in 1572, a braver man stepped up. An engineer named Rafael Bombelli decided to stop panicking and simply *use* these impossible square roots, following ordinary rules of arithmetic, to see what happened. And what happened was miraculous: when he carried them through the calculation, the impossible numbers cancelled each other out at the end and left behind perfectly ordinary, correct, real answers. The ghosts were doing real work. They were a bridge you could walk across to reach solid ground on the other side, even though the bridge itself seemed to be made of fog.

This should have been the moment everyone celebrated. Instead, the discomfort lasted for centuries. The numbers worked, but nobody could say what they *were*. And then, in 1637, the great philosopher René Descartes gave them a name. He did not mean it kindly. He called them "imaginary," to mark them as fake, fictitious, things that existed only in the mind and not in reality.

The insult stuck. It is still stuck today. We still call them imaginary numbers, four hundred years after a man who did not believe in them sneered at them in a footnote.

Do you see it? It is the same disease from the very first page of this book. A name, attached carelessly, that turned out to be wrong, and that outlived the truth simply because it was repeated often enough. Einstein never said the quote. Pythagoras did not discover the theorem. And the "imaginary" numbers are not imaginary. The label was the only lie.

It took Leonhard Euler, the hero of this chapter, to start dragging these numbers into the daylight in the 1700s, using them fearlessly and giving them the simple symbol we still use: the letter i. And it took Carl Friedrich Gauss, a little later, to finally show everyone where i actually lives, and to give the numbers a fairer name: *complex* numbers. Not imaginary. Just complex.

So where does i live? What is it, really? That is the question Descartes could not answer and Gauss could. And the answer is so simple that once you see it, the three hundred years of fear start to look a little silly.

i Is Just a Turn

Forget square roots for a moment. Picture all the ordinary numbers laid out along a straight road. Zero in the middle. The positive numbers run off to the right. The negative numbers run off to the left. You are standing on the number 1, over on the right.

Now multiply your position by minus one. You land on minus 1, on the far left. Multiplying by minus one flipped you straight across to the opposite side of zero. Think about what that flip *is,* as a movement. It is a half turn. A U-turn. A rotation of 180 degrees, swinging you from facing right to facing left.

Now here is the single question that unlocks everything, the question that dissolves three centuries of confusion in one stroke.

Multiplying by minus one turns you 180 degrees. So what operation, done *twice,* would add up to that same 180-degree turn?

The answer is obvious the moment you ask it. Half of a 180-degree turn is a 90-degree turn. Turn 90 degrees, then turn 90 degrees again, and you have made the half circle. So there must be some operation meaning "turn a quarter circle," which, applied twice, equals multiplying by minus one.

That operation is i. That is the entire secret. i is the instruction: turn ninety degrees. And "doing i twice equals minus one" is just "two quarter turns make a half turn." Written in symbols, that is the famous, frightening definition that scared everyone for centuries: i times i equals minus one. All it ever meant was a quarter turn, done twice.

But now think about what a quarter turn does to you, standing on that number road. You were facing along the line, left and right. Turn 90 degrees, and you are now facing *off* the road entirely, into open ground that the old numbers never used. i does not live on the number line. It points into a second dimension, a whole new direction sitting at right angles to the ordinary numbers, that was there the entire time and nobody had looked.

This is what Gauss finally made everyone see. The numbers are not a single road. They are a flat plane, a whole open field. The ordinary numbers run east-west. The so-called imaginary numbers run north-south. And i is simply the step that takes you from one direction to the other. The steering wheel of space.

So i is not a ghost, and it is not imaginary. It is the most concrete thing imaginable: a quarter turn. That is stranger number five, and it was never a ghost. It was a direction nobody had named.

Growth, Steered Sideways

Now we put two of our strangers together, e and i, and this is where the whole thing catches fire.

We have e, which is pure growth. Normally, raising e to a larger and larger power means more and more growth, racing straight out along the number line toward infinity. That is what unchecked growth does. It runs away.

But what happens if we feed i into that growth? Remember what i does: it turns you ninety degrees. So now we are asking for growth whose direction is constantly being twisted sideways, a quarter turn, away from wherever it is currently heading.

Picture it carefully. You are standing on the number

1. Your growth engine wants to push you to the right, outward, bigger. But i grabs that push and rotates it ninety degrees, so instead of stepping right, you step *up*. Now you are in a slightly new place, facing a slightly new way. Your engine fires again, trying to push you straight ahead, and again i twists it ninety degrees sideways. And again. And again, every instant, forever.

You have felt this exact thing with your own arm. Tie a rock to a string and swing it around your head. The rock is desperate to fly off in a straight line, but the string keeps yanking it inward, at a right angle to its motion, again and again, every instant. The rock cannot escape outward, because something keeps turning it. And it cannot stop, because it still has all its speed. Trapped between those two facts, it does the only thing left to do. It travels in a perfect circle, around and around, at a steady pace, forever.

That is precisely what growth steered by i does. It cannot run away to infinity, because it is never allowed to point "away" for more than an instant before being turned. It cannot stop, because it is still growth.

Caught between those two rules, it walks in a circle.

So here is the payoff, one of the most beautiful turns in all of mathematics. e raised to an *imaginary* power is not some monstrous exploding number. It is just a position on a circle. The bigger the power you feed it, the further around the circle you have walked. Growth, bent into a ring.

We can write that down, and this, not the famous one-liner, is the real jewel of the chapter:

$$e^{i\theta} = \cos\theta + i\sin\theta$$

Do not let the symbols frighten you, because the sentence underneath them is gentle. The little θ (say "theta") is just a measure of how far around the circle you have walked. And cos and sin, those two old words from school trigonometry that terrified everyone, are nothing more than the two coordinates of where you end up: how far across you are, and how far up you are. That is all they are. Positions on a circle.

So read the whole formula in plain words: *e raised to*

an imaginary power is simply the point you arrive at by walking that far around a circle. Growth and turning, married, tracing a perfect ring. Three of our strangers, e and i and (soon) π, were never strangers at all. They were three descriptions of the same circle.

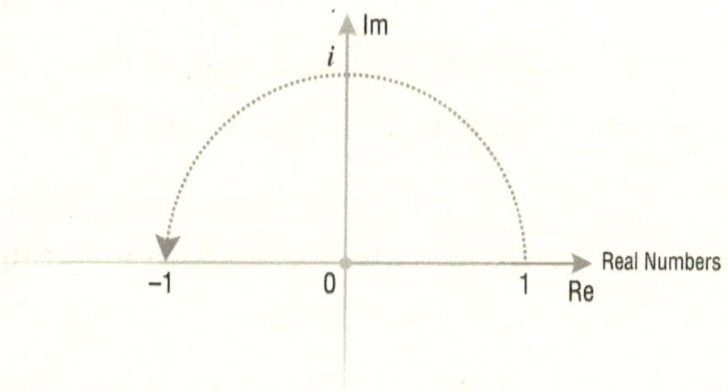

Figure 4: Figure 4.1 – Euler's identity on the unit circle: walking an imaginary angle traces a path through cos and sin.

A Look Under the Bonnet

You can skip this section. The book will still make complete sense without it, and the circle picture you already have is true. But this is a book about checking things for yourself, and some readers will want to see, with their own eyes, that the circle really is exact and not just a pretty story. So here, gently, is the machine that proves it. Come along if you like. Step over it if you would rather.

It turns out that many mathematical functions can be written as an endless recipe: an infinite string of simpler and simpler terms, added up, that get smaller and smaller so the total settles on a real answer. Growth has such a recipe. So does "how far across the circle" (cos). So does "how far up the circle" (sin). For three hundred years these three recipes were known, sitting in separate drawers, looking nothing like one another.

The recipe for growth, e raised to some amount x, runs like this:

> $1 + x + \frac{x^2}{2} + \frac{x^3}{6} + \frac{x^4}{24} + \cdots$ (each term divides by a bigger number, so the tail shrinks away)

The recipe for "how far across" (cos) runs like this:

> $1 - \frac{\theta^2}{2} + \frac{\theta^4}{24} - \cdots$ (notice: only the even steps, and the signs flip, plus, minus, plus)

And the recipe for "how far up" (sin) runs like this:

$\theta - \frac{\theta^3}{6} + \frac{\theta^5}{120} - \cdots$ (only the odd steps, signs flipping again)

Now we do the daring thing. We take the recipe for growth, and instead of feeding it an ordinary amount, we feed it an imaginary one: $i\theta$. The growth recipe does not care; it just follows its rule. But watch what happens, because i has a hidden rhythm. When you multiply i by itself over and over, it cycles in a loop of four: i, then minus 1, then minus i, then back to 1, then i again, forever. A four-beat drum.

When that four-beat rhythm marches through the growth recipe, it sorts the endless string of terms into two kinds: terms with an i stuck to them, and terms without. Pull them gently into two separate piles. And here is the moment three centuries had been waiting for. The pile *without* the i reads, term for term, sign for sign: $1 - \frac{\theta^2}{2} + \frac{\theta^4}{24} - \cdots$ which is *exactly* the recipe for "how far across." And the pile *with* the i reads: $\theta - \frac{\theta^3}{6} + \frac{\theta^5}{120} - \cdots$ which is *exactly* the recipe for "how far up."

Nothing was forced. Nothing was left over. The growth recipe, fed an imaginary number, fell apart on its own into the two coordinates of a circle. The picture was not poetry. It was the literal truth, hiding inside the algebra the whole time, waiting for someone to feed it an imaginary number and watch it bloom.

That is the proof. e steered by i *is* the circle, exactly. Now back to the strangers.

Halfway Around

We are nearly home. We have $e^{i\theta}$ meaning "start on the number 1, and walk θ around a circle of size 1." We have only one choice left to make: how far to walk.

We choose π.

Why π? Because π is the number that measures the circle, and the cleanest way to measure a journey around a circle is in lengths of its own radius. Lay the radius along the rim, and it turns out that going the whole way around uses exactly 2π of them. The full lap is 2π. Which means that walking just π is walking exactly *half* the circle. A half lap. A semicircle.

So picture it, one final time. You stand on the number 1, at the right-hand edge of the circle. You walk halfway around. Where do you end up? Directly opposite. On the far left-hand edge. And the left-hand edge of a circle of size 1 is the number minus 1.

That is the whole of it:

$$e^{i\pi} = -1$$

It looks like a monster carved from three of the most intimidating symbols in mathematics. It is, in truth, a gentle stroll: "begin at 1, walk half a lap around a circle, arrive at minus 1." A child could trace the journey with one finger. We will, at the end of this chapter, with a pencil.

Bringing It Home to Zero

We have landed on minus 1. We are one small step from nothing at all.

Take the line we just earned, $e^{i\pi} = -1$, and add one

to both sides. On the right, minus 1 plus 1 becomes 0. And we are left holding the most famous short sentence in mathematics:

$$e^{i\pi} + 1 = 0$$

There it is. The half lap that lands on minus 1, plus a single step of one, brings you back, exactly and perfectly, to zero. The circle of the whole idea closes on nothing, where it began.

And look who is standing in that one short line, holding hands. Zero, the nothing that India had to be brave enough to invent. One, the first something, the seed of all counting. π, the endless number from the roundest shape there is. e, the ceiling of all growth, found in a banker's ledger. And i, the quarter turn that three hundred years of clever people called impossible. Five strangers, from five different worlds, in one sentence, balancing to exactly zero. Not roughly. Not nearly. Not 0.0001. Exactly zero, forever.

Discovered, Not Invented

Now stop and feel the full size of what just happened, because this is the part that keeps people up at night.

Five numbers, found by five different kinds of people who never met, separated by oceans and centuries, each one chasing something completely different. A scribe measuring a wheel. A banker counting coins. A duelling Italian wrestling with a cubic. The plain old ideas of nothing and something. Not one of them was looking for the others. Not one of them had any reason to fit with the rest.

And they fit together perfectly. Exactly. On the first try. Forever.

This forces a very old and very deep question, maybe the deepest question in all of mathematics: did we *invent* this, or did we *discover* it?

When you build a chair, you invented it. It would not exist without you, and you could have made it differently. But when an explorer reaches a new mountain, they did not invent the mountain. It was always there. They merely walked far enough to find it.

Which is mathematics? If we invented it, freely, as a human game, then there is no reason on earth these five numbers should agree. We would simply have made them up separately and moved on. But they do agree, exactly, whether we like it or not, whether we expected it or not. Nobody chose for $e^{i\pi} + 1$ to equal zero. It simply does, and it would still be true if no human had ever been born to write it down.

That points to something unsettling and wonderful. We did not invent this equation. We *found* it. The five strangers were always family. They were sitting in the same house the entire time. We just kept walking into different rooms, meeting one of them at a time, never realising they all shared a kitchen. Euler's identity is the moment somebody finally opened all the doors at once and saw them together.

The physicist Eugene Wigner once wrote a famous essay marvelling at what he called the unreasonable effectiveness of mathematics: the eerie way that maths, dreamed up by humans for its own sake, keeps turning out to describe the real universe with perfect accuracy, as if it were the language reality was written in all along. This equation is the purest taste of that mystery. It is the universe quietly admitting that, underneath everything, it is all of one piece. Which is the same promise this book made you on its first page: that the world, against all odds, is built to be understood.

That is why an equation that builds nothing matters more than almost any equation that does.

The Name on the Door

We call this Euler's identity, after Leonhard Euler, and here, for once, the name on the door belongs to someone who earned it. Euler really did find the engine: the formula about walking around the circle, $e^{i\theta} = \cos\theta + i\sin\theta$. He is the one who pulled the "imaginary" numbers into the daylight. He is even the one who gave them the symbol i, and gave growth its symbol e. Look back at the equation: two of its five characters are wearing notation that Euler himself invented. He did not just discover the relationship. He handed us the very alphabet we wrote it in.

But there is one small, careful thing to flag, because this is a book about small, careful things.

You have probably heard that the great physicist Richard Feynman called this "the most remarkable formula in mathematics," and "our jewel." He did say exactly that. It is a real quote, from his famous lectures, and we can trust it completely, unlike a certain quote about a fish.

But read it precisely. Feynman said it about the *engine*: the full formula about the circle, $e^{i\theta} = \cos\theta + i\sin\theta$. He did not say it about the tidy one-line gem, $e^{i\pi} + 1 = 0$, that ends up framed on walls and printed on mugs. That gem is just the engine, frozen at a single instant, half a lap around. It is genuinely beautiful. But the thing Feynman actually called the jewel was the whole turning machine, not one photograph of it.

People quote him praising the poster. He was praising the engine that printed the poster. A tiny difference. But getting tiny differences right is the entire reason this book exists.

The Man Who Sped Up

It is worth pausing on Euler himself, because he is one of the most extraordinary humans who ever lived, and almost nobody outside mathematics knows his name.

He was born in Basel, Switzerland, in 1707, the son of a pastor who expected him to follow into the church. Instead he became, by a wide margin, the most prolific mathematician in all of history. He wrote somewhere around 866 papers and books, on so vast a range of subjects that his collected works fill roughly eighty large volumes. Editors were still publishing his unpublished manuscripts for decades after his death, working through the backlog he left behind. There is barely a corner of mathematics he did not touch and reshape, and much of the notation in every textbook today is simply his handwriting, frozen in place.

But here is the part that turns admiration into something closer to awe. In his later years, Euler went blind. First one eye, then, by his sixties, both. For most people that would be the end of a career built on reading symbols and staring at diagrams.

It was not the end of his. If anything, his output *increased.* He had a memory so prodigious that he could hold enormous calculations entirely in his head, turning them over in the dark and dictating the results to assistants who wrote them down. In one stretch he is said to have produced, on average, a finished mathematical paper every week. Blind. From memory. When a fire destroyed much of his work, he reconstructed it from his mind. He is reported to have remarked that without his sight he would have fewer distractions to take him away from his mathematics.

He worked until his final day. The story goes that he died in the middle of a calculation, mid-thought, surrounded by family. A man who could not see, who could "see" further into the structure of reality than almost anyone before or since.

When you draw the circle at the end of this chapter, you are retracing the path of a man who could not see the circle, and understood it more deeply than the people who could.

The Engine Runs Everything

And now the final twist, the one that drags this useless, beautiful equation crashing back into the real, working world.

The gem builds nothing. But the engine behind it, Euler's formula, the one about going around a circle, is quietly running almost every machine you own.

Think again about what $e^{i\theta}$ actually describes: something going around and around a circle, forever, at a steady pace. Now ask yourself what that *is*, physically. It is anything that repeats. A cycle. A vibration. A wave. And the moment you can describe a wave, you can describe almost everything that carries energy or information through the modern world.

The alternating current humming in the wires of your wall is a wave, and electrical engineers describe it using exactly this mathematics; in fact they lean on i so constantly that, to avoid confusion with the symbol for electric current, they rename it j and carry on. Every radio signal is a wave. Your wifi, your mobile phone call, every song you stream, every photograph that crosses an ocean of cable to reach your screen, every colour of light your eye catches, all of them are waves, and all of them are handled, by the people who build the networks, with this exact tool.

There is an even deeper magic here, discovered after Euler, called the Fourier idea. It says that *any* signal at all, no matter how complicated, no matter how jagged and messy, can be broken apart into a sum of simple, pure circular motions, each one an $e^{i\theta}$ turning at its own speed. A chord of music is a handful of these circles spinning together. The shape of your voice down a phone line is a crowd of them. A photograph, an MP3, a video call, the noise-cancelling in a pair of headphones, the entire field of digital signal processing, all of it works by splitting the world into Euler's spinning circles, doing something to them, and adding them back together.

So the most "useless" equation in this book is a single frozen frame of the single most useful tool in engineering. The gem on the wall and the engine in the machine are the same object, seen from two sides. Beauty and usefulness, which everyone treats as opposites, turned out to be family too. They always were.

So Check It

You cannot cut this one out of a cereal box. But you can still be the judge, with nothing more than a round object and a pencil.

Find anything circular: a cup, a coin, a roll of tape. Trace around it to draw a circle. Mark the centre with a dot, and mark the point on the far right edge of the rim. That right-hand point is your starting line, the number 1.

Now put your fingertip on that point and slide it, slowly, halfway around the rim, until you reach the point directly opposite, on the far left edge.

That point is minus 1. You have just walked $e^{i\pi}$ with your own hand. And if you step from there back in toward the centre by one unit, you arrive at the middle: zero. $e^{i\pi} + 1 = 0$, traced on the back of an envelope, in the time it takes to finish a cup of tea.

You did not need to believe Euler, who could not see

the circle he understood. You did not need to believe Feynman, or Descartes, who got the name wrong, or me.

Draw the circle. Walk halfway round.

You will land on the other side. You always will. And five strangers, who never should have met, will be standing there together, exactly where the equation promised they would be.

There is a temptation, after an equation like this, to imagine that mathematics lives in some perfect crystal palace, hovering far above the grubby physical world, too pure ever to get its hands dirty.

Euler knew better. The same man who balanced five strangers on the head of a pin spent much of his life on the least heavenly subjects imaginable: the roll of a ship, the load on a beam, the flow of water through a pipe. He wrote some of the very first equations describing how fluids move, the founding mathematics of running water and moving air. To him, the crystal palace and the water pipe were the same building.

And he had learned his craft, as a boy in Basel, from

a family of mathematicians as dazzling as they were poisonous to one another: the Bernoullis. We are about to meet them, because one of them, Daniel, found a principle about moving fluids so useful that engineers still reach for it every single day, and so slippery that almost the entire world, teachers and pilots and encyclopaedias alike, would seize upon it to explain how an aeroplane flies, and get the explanation gloriously, confidently wrong.

So we climb down out of the crystal palace and step back into the open air, where the wind is moving and nothing holds still. And we begin with a question you have certainly been handed an answer to already, an answer repeated almost everywhere, that happens not to be true.

Chapter Five

Bernoulli's Equation

The Less It Pushes

$$P + \frac{1}{2}\rho v^2 + \rho g h = \text{constant}$$

Ask almost anyone why an aeroplane stays up, and you will hear the same story. It is taught in schools. It is printed in encyclopaedias. It is repeated by pilots, by physics teachers, by clever people at dinner parties. It goes like this.

A wing is curved on top and flat underneath. So the air going over the top has a longer way to travel than the air going underneath. The two streams split at the front of the wing and have to meet up again at the back, so the air on top must go faster to cover the extra distance in the same time. And faster-moving air, by something called Bernoulli's principle, has lower pressure. So there is less pressure pushing down on top of the wing than pushing up underneath it, and that difference lifts the plane.

It is a tidy story. It uses a real principle. It is told by experts. And it is wrong.

Not slightly wrong. Wrong at its heart, in a way that NASA itself has had to publish corrections about, because the false version is so widespread. There is no rule of physics that says the two streams of air must meet up again at the back of the wing. They do not. The air over the top actually arrives at the back *sooner* than the air underneath, and it moves faster than the "meeting up" story could ever explain. The whole tale rests on a coincidence that simply is not true.

And here is the part that makes this a perfect chapter for this book. The principle the story leans on, Bernoulli's principle, is completely, beautifully real. The equation is one of the most useful in all of engineering. It is only the famous *story about flight* that is the myth. So I am going to do two things. First, we will meet the real principle, which is true and lovely and hiding in a hundred everyday places. Then we will come back to the wing and find out how flight honestly works, which turns out to be even better than the fairy tale.

The Real Idea

Strip away the wing for a moment and here is the actual principle, and it is genuinely surprising:

When a fluid speeds up, its pressure drops.

By "fluid" we just mean anything that flows, air or water or any gas or liquid. And the claim is that the faster it is moving, the *less* it presses sideways on its surroundings. This feels backwards. Fast-moving water from a hose hits you hard, so surely fast means more push, not less? But the hard hit is the water being *stopped;* that is a different thing. While the fluid is smoothly flowing, the faster it flows, the gentler it presses on the walls around it. Speed and pressure trade off against each other. You can have one or the other, but not both at once.

Why? The reason is one we have met before in this book: energy is conserved. A flowing fluid carries a fixed budget of energy, and it can hold that energy in different forms. It can hold it as pressure, the push it exerts. It can hold it as motion, its speed. It can hold it as height, how high it has been lifted. The grand bookkeeping is the equation at the head of this chapter:

$$P + \frac{1}{2}\rho v^2 + \rho g h = \text{constant}$$

Do not be put off. Read it as a simple budget with three pockets. The first term, P, is the pressure, the push. The middle term is the energy of motion: ρ (say "rho") is just how dense the fluid is, how much stuff is packed into it, and v is its speed. The last term is the energy of height, with g the strength of gravity and h the height. And the whole point is the words on the right: the total stays *constant*. The three pockets must always add up to the same amount.

So watch what happens when the fluid speeds up along a level pipe, where the height does not change.

The middle pocket, the energy of motion, swells. The total cannot change. So one of the other pockets must shrink to pay for it, and since the height is fixed, the only one left to give is the pressure. The pressure drops. The fluid spent some of its push to buy speed.

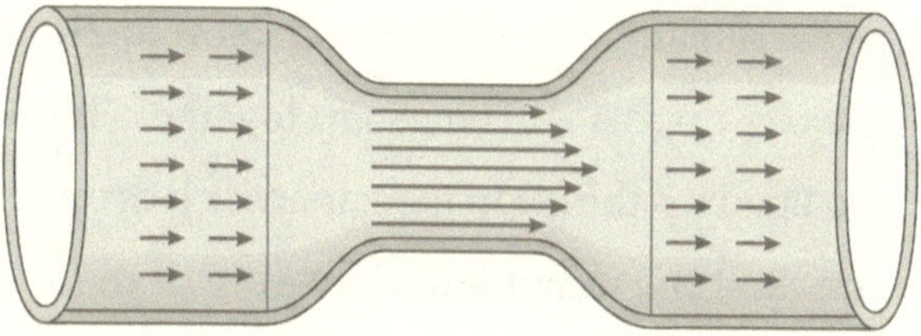

Figure 5: Figure 5.1 — Bernoulli's principle: as fluid speeds up through a constriction, its pressure drops.

That is the whole of Bernoulli's principle. A fluid has a fixed allowance, and every bit of speed it gains is paid for out of its pressure. Faster flow, lower pressure. Always.

You Can Feel It

This is not an abstraction. You can watch it work in the next sixty seconds with two sheets of paper.

Hold them up in front of your face, hanging down, parallel, a few centimetres apart, like two flags side by side. Now blow a stream of air *between* them. Common sense screams that they should blow apart. They do the exact opposite. They snap together. By blowing, you sped up the air in the gap, which dropped its pressure, and the now-higher pressure of the still air on the outside pushed the two sheets together. You made a low-pressure zone with your own breath and the sheets fell into it.

Once you have seen it, you start to notice it everywhere. In a storm, it is not really the wind pushing a roof in that tears it off; it is the wind racing *over* the top of the roof, dropping the pressure up there, while the ordinary still air trapped inside the house pushes up, and the roof is sucked off from above. A perfume atomiser or an old paint sprayer works by blowing a fast jet of air across the top of a tube standing in the liquid; the fast air is low pressure, so the higher pressure on the liquid below shoves it up the tube and into the jet, where it shatters into a fine mist. Two large lorries side by side on a motorway feel a gentle pull toward each other, because the air squeezed between them is moving faster than the air on their outer sides. Even your own arteries obey it: where blood speeds up through a narrowing, its pressure falls, which doctors must account for when they read the flow through a clogged or bulging vessel.

The principle is real, it is everywhere, and it is genuinely useful. Which makes it all the more important to be honest about the one place it gets misused.

The Story You Were Told

So let us go back to the wing, and look closely at the famous explanation, because it is a textbook example of how a true principle can be welded onto a false story until the whole thing looks solid.

The story has one fatal assumption, and you can spot it now that you are looking. It claims that the air going over the top and the air going under the bottom, having split apart at the front edge of the wing, *must rejoin at the back edge at the same moment.* From that single assumption everything else follows: if the top path is longer and the trip must take equal time, the top air is forced to hurry, and Bernoulli does the rest.

But why on earth would the two streams have to meet up again? They were never partners. Nothing ties a parcel of air going over the top to a particular parcel going underneath. They are not holding hands. They have no appointment to keep at the back of the wing. The assumption was simply invented because it makes the arithmetic work, and then quietly forgotten, so that the conclusion looks like physics when it is really just a guess wearing a lab coat.

And when we actually measure the air, the guess collapses. The air going over the top of a real wing moves *much* faster than the "meet up at the back" idea requires. In fact it gets to the trailing edge well before the air underneath. The streams never rejoin as partners at all. The premise is not just unproven. It is false.

There is a quick way to prove the whole story cannot be the real explanation. If lift came from the top surface being longer than the bottom, then a wing with a flat top and bottom could not lift, a perfectly symmetric wing could not lift, and no aircraft could ever fly upside down. Yet flat wings lift, symmetric wings lift, and aerobatic pilots fly inverted all afternoon. A paper aeroplane, flat as a envelope, flies. The story does not merely have a hole. It is contradicted by every airshow on Earth.

How a Wing Really Works

So throw the fairy tale away, and let us see what is actually happening, because the truth is simpler and more satisfying.

Watch the air *behind* a flying wing, and you see the real story at once. The wing leaves a great river of air flung downward in its wake. Smoke trails in a wind tunnel show it plainly; the air arrives moving roughly level, and leaves angled down toward the ground. The wing has grabbed a vast amount of air every second and thrown it downward.

And now you already know the rest, because we proved it three chapters ago. Forces come in pairs. If the wing pushes that air *down,* then by Newton's third law the air pushes the wing *up* with exactly equal force. That upward shove is lift. That is it. A wing flies for the very same reason a rocket flies and a person on a skateboard rolls backward when they throw something: throw mass one way, get pushed the other. A wing is a machine for throwing air downward, and the air, in return, holds it up.

This is why a flat wing works, as long as it is tilted so its leading edge rides a little higher than its trailing edge, the angle pilots call the angle of attack. Tilt the wing, and it deflects the oncoming air downward, and up it goes. Tilt it more, and it throws more air down, and lift increases, until you tilt too far, the smooth flow breaks away, and the wing stalls. It is why a symmetric wing flies, and why a plane can fly upside down: point the thing so it drives air downward, by whatever shape and angle, and you will get lift.

Now, here is the lovely reconciliation, the part that brings Bernoulli back in from the cold. The pressure story and the air-throwing story are not rivals. They are two descriptions of the very same event, seen from two sides. The air really does move faster over the top of the wing, and so, by Bernoulli's perfectly correct principle, the pressure on top really is lower, and that low pressure really does help suck the wing upward. Both the faster flow over the top *and* the downward-flung wake are part of one connected motion: the wing bending the whole flow of air around it and downward. You can add up the pressures all over the wing's surface, or you can measure the air thrown downward behind it, and you will get the very same lift, to the same number. They must agree, because they are the same physics counted two ways.

So Bernoulli was never the villain. The villain was a made-up rule about air meeting at the back. The honest answer keeps Bernoulli, adds Newton, and admits one last humbling thing: the deepest question of all, *why* the air chooses to hug the curved top of the wing and follow it downward instead of flying straight off into space, turns out to be genuinely subtle, tangled up in the stickiness of real air, and is something even good engineers describe carefully rather than glibly. We can predict lift to staggering precision. We can build wings that carry five hundred people across an ocean. And we can still be honest that the full, first-principles "why" of how the flow bends is deeper than a single slogan. That honesty is not a weakness. It is the difference between understanding flight and reciting a story about it.

Bernoulli Everywhere

Step away from wings and the real principle turns up all over the built world, quietly doing its job without any myth attached.

It is how a carburettor in an old engine drew fuel into the air stream, and how the same trick measures the flow of gas and liquid in pipes across industry: narrow the pipe, the fluid speeds up, the pressure drops in the narrow section, and from how far it drops you can read the speed of the flow. It is at work in the hiss of a spray can and the draw of a chimney. It shapes how a sailing boat is driven by its sails, which act as upright wings and let a skilled sailor cut across the wind, even partly into it. It helps explain why a spinning ball curves through the air, dragging the air around faster on one side than the other. It is in the medicine that reads the blood through a narrowed heart valve. Daniel Bernoulli's principle is one of the genuine workhorses of engineering, holding up far more of the modern world than the one flashy job it keeps getting wrongly credited for.

The Name on the Door

The principle is named after Daniel Bernoulli, who set it out in a book on the motion of fluids in 1738. And the story behind the name is one of the most spectacularly bitter in the whole history of science, because the Bernoullis were not just one mathematician. They were a dynasty of them, a single family that produced a startling run of brilliant mathematicians across three generations, and they competed with each other like sworn enemies.

The worst of it was between Daniel and his own father, Johann Bernoulli, himself one of the finest mathematicians in Europe. When Daniel produced his great work on fluids, his father was consumed by jealousy.

Johann wrote a rival book on much the same subject, and then did something breathtaking: he had it dated to an earlier year than it was truly written, so that it would appear to have come *before* his son's, and so that the world would credit the father with the son's ideas. It was a deliberate forgery of history, by a father, to rob his own child of the glory. By some accounts Johann had earlier thrown Daniel out of the house after the two were named to share a prize Johann felt he alone deserved.

So even here, at the polite-sounding name above the door, we find the usual human mess: envy, theft, a faked date, a father and son who could not stand to share. And there is one more thread worth pulling, because it ties this chapter to another. Johann Bernoulli, the jealous father, was also the teacher of a quiet young man in the same Swiss town, a student he recognised as something rare and tutored privately on Saturday afternoons. That student was Leonhard Euler, whom we will meet two chapters from now, and who would go on to outshine the entire feuding clan. The Bernoullis fought tooth and nail over who owned which idea. The ideas, as ever, did not care. The pressure still drops when the fluid speeds up, whoever dated their book whenever.

Is It Really True?

Completely. This is one of the most thoroughly confirmed and relied-upon equations in engineering, checked in pipes and wind tunnels and pumps every day for nearly three centuries. And planes, let us be very clear, absolutely do fly. The myth in this chapter was never that flight is fake. The myth was only in the *explanation,* the comforting little story about air meeting at the back of the wing. The flying is real. The principle is real. It was just the bedtime story joining them that was invented.

Like every law in this book, Bernoulli's equation also knows its borders, and an honest engineer respects them. It assumes a smooth, steady, tidy flow, and it leaves out the stickiness of real fluids, the friction that drags and the turbulence that churns. Push into fast, messy, compressible flows, near the speed of sound, and you need more than this clean equation can give. But within its proper home, the everyday flow of everyday fluids, it is exact, dependable, and quietly everywhere.

So Blow on a Page

You do not need to take the schoolbook's word for how wings work. You certainly do not need the fairy tale. You can hold the real principle between your fingers right now.

Tear a strip of paper, or grab a full sheet. Hold one edge just below your bottom lip, letting the rest droop sadly down and away from you. Now blow steadily straight out, across the top of the drooping page. It will rise and flutter up toward your stream of air, as if pulled by an invisible hand. You sped up the air across the top, the pressure there dropped, and the ordinary still air beneath pushed the paper up into the gap. That is Bernoulli, in your own breath, on a scrap of paper, and you can do it again and again before bed.

And notice what just happened to the famous flight story. You did the thing it describes, fast air over a curved-down top giving lift, with a single sheet of paper, no second surface underneath, no streams meeting at any back edge. The real principle held up the paper exactly as it should, and the fairy tale was never needed at all. Reality kept its appointment. The story never had one.

We have now spent five chapters among solid, touchable things: corners, carts, falling apples, flowing air. Things you can push, drop, and feel. But the next great equation is about something you cannot touch at all, and were probably told is simply empty: the space around you. We are about to discover that even where there is no air, no water, nothing whatsoever to push, space itself is quietly humming with invisible fields. And when those fields are made to ripple, the ripple racing toward your eyes at the fastest speed there is, that ripple is light.

Chapter Six

Maxwell's Equations

What Light Is

$$\nabla \cdot \mathbf{E} = \frac{\rho}{\varepsilon_0} \qquad \nabla \cdot \mathbf{B} = 0$$

$$\nabla \times \mathbf{E} = -\frac{\partial \mathbf{B}}{\partial t} \qquad \nabla \times \mathbf{B} = \mu_0 \mathrm{J} + \mu_0 \varepsilon_0 \frac{\partial \mathbf{E}}{\partial t}$$

Right now, in the space your body is sitting in, the air looks empty. It is not.

Passing through you, this very instant, are the radio broadcasts of a dozen stations, the wifi of your home and your neighbours', the chatter of distant phones, the faint warmth of the Sun on your skin, the glow of every lamp in the room, and a whisper of light left over from the birth of the universe itself. They are streaming through the chair, the walls, and you, all at once, all the time. You feel almost none of it.

And here is the secret this chapter is going to give you. Every single one of those things is the same thing. The radio and the wifi and the sunlight and the lamplight and the ancient afterglow of the Big Bang are not cousins. They are not even close relatives. They are literally, exactly the same phenomenon, differing only in one number, like notes at different pitches played on the same string. To understand what that one thing *is*, four short equations had to be written, and when they were, a man sitting quietly with a pen discovered the answer to a question humanity had been asking since the first eye opened: what is light?

He found out on paper. Before anyone had ever measured it. He worked out what light is the way you might work out the answer to a sum, and he was right. It remains, to this day, perhaps the single most astonishing thing anyone has ever done with a pencil.

To get there, we have to start with two mysteries that, for thousands of years, nobody dreamed had anything to do with each other.

Two Mysteries

The first mystery was electricity. The ancients knew that a piece of amber, rubbed with fur, would crackle and draw bits of straw toward it. They knew that lightning could split a tree. But electricity was a curiosity, a party trick and a terror, with no apparent rules.

The second mystery was magnetism. Certain rocks, lodestones, would cling to iron. A sliver of such rock, floated on water, would swing to point north, and that single strange fact, packed into a compass, let sailors cross oceans they could not see across. But magnetism, too, seemed its own separate riddle.

Two unrelated oddities. One about sparks and storms, the other about rocks and compass needles. For most of human history nobody had the slightest reason to connect them. They sat in different drawers of the world. And then, over a few decades in the 1800s, a handful of people pulled the two drawers open at the same time and found that they were joined at the back.

The First Clue

In 1820, a Danish teacher named Hans Christian Oersted was giving a lecture, with a wire carrying an electric current lying near a compass. He noticed something that should have been impossible by the thinking of the day. Whenever the current flowed, the compass needle twitched and turned, as though a magnet had been waved near it. There was no magnet. There was only electricity moving through a wire.

Electricity, when it moved, made magnetism.

This was a thunderclap. The two mysteries were not separate after all. A flowing current, pure electricity, could push a compass needle, pure magnetism, as surely as any lodestone. People raced to explore it, and a decade later the other half of the secret fell into place, found by one of the most remarkable people in the history of science.

Michael Faraday had been born poor, the son of a

blacksmith, and was apprenticed to a bookbinder as a boy. He had almost no formal schooling and, by his own cheerful admission, knew next to no mathematics. What he had instead was a pair of the finest experimental hands that ever lived and an imagination that could *see* into the invisible. And Faraday asked the natural question. If moving electricity makes magnetism, can moving magnetism make electricity?

It can. In 1831 he pushed a magnet in and out of a coil of wire, and each time it moved, a current flickered to life in the coil, out of nothing, with no battery anywhere. Move a magnet near a wire, and you make electricity. That single discovery is the reason your home has power. Every power station on Earth, whether driven by steam, falling water, or wind, is in the end just a machine for shoving magnets past coils of wire to make exactly the current Faraday first coaxed out in his laboratory.

So now the picture was symmetric and strange. Changing electricity makes magnetism. Changing magnetism makes electricity. Each can give birth to the other, but only while it is *changing*. Hold them still and the spell breaks. The two mysteries were not just related. They were forever turning into one another.

The Bookbinder's Apprentice

Michael Faraday is worth stopping on, because his life is one of the most improbable in all of science, and because almost no one who flicks a light switch has ever heard his name.

He was born in 1791 into real poverty, the son of a London blacksmith, in a family that did not always have enough to eat; there is a story of the boy being handed a single loaf of bread and told to make it last a week. There was no money for proper schooling. At fourteen he was apprenticed to a bookbinder, and that turned out to be the great stroke of luck in his life, because a boy who binds books can also read them. Faraday read everything that passed through the shop, and the science books above all, teaching himself the subject that would consume the rest of his days, one bound volume at a time.

He got through the door by sheer nerve. After attending public lectures by the famous chemist Humphry Davy, Faraday wrote the lectures up in careful detail, bound them beautifully, the one craft he truly had, and sent the volume to Davy as a kind of silent application. Davy, impressed, and soon needing an assistant, took the young man on, and so the bookbinder's apprentice walked into the Royal Institution, where he would remain for the rest of his life, and where he would in time entirely eclipse his famous master.

What he did there reads like a table of contents for the modern world. He discovered electromagnetic induction, the principle inside every generator. He built the first primitive electric motor. He laid down the basic laws of electrochemistry. He even gave us much of the vocabulary we still use, including the words "electrode," "ion," and "field." And he did nearly all of it with little more than superb hands, bottomless curiosity, and that strange gift for *seeing* the invisible.

What makes him easy to love is that none of the success changed him. He stayed modest and decent to the end. He turned down a knighthood, preferring to remain plain Mr Faraday. He twice refused to become president of the Royal Society. When the British government asked him, as an expert on gases, to help develop poison gas for the Crimean War, he refused outright, calling it an atrocity, though he admitted he had no doubt it could be done. And when he was offered burial among the nation's heroes in Westminster Abbey, he declined even that, choosing a simple grave instead. The man who did more than almost anyone to electrify the world wanted, in the end, to be ordinary.

This is the person who handed Maxwell the central clue, and the gulf between the two men is part of what makes the story so moving. Faraday was poor, self-taught, and innocent of mathematics. Maxwell was a comfortable, university-trained gentleman fluent in the most advanced mathematics of the age. By rights they should have had nothing to say to one another. Instead Maxwell read Faraday's work, recognised the deep genius buried in those plain picture-thoughts about lines and fields, and set himself to give them a mathematical body, always insisting that he had merely translated Faraday's ideas, never replaced them. The labourer's son saw the truth. The gentleman wrote it down. Neither could have done it alone.

Faraday's Invisible Lines

Faraday, who could not do the equations, did something arguably deeper. He came up with a picture, and it turned out to be one of the most important ideas in all of physics.

You have seen it, probably as a child. Scatter iron filings on a sheet of paper and hold a magnet underneath, and the filings leap into a pattern, looping curves arcing from one end of the magnet to the other, tracing invisible lines through the empty space around it. Faraday looked at those lines and made a daring leap. What if they are *real*? What if the space around a magnet, or around an electric charge, is not empty at all, but is filled with a genuine, physical *something*, a condition stretched through space, that the filings are merely revealing?

He called it a field. The magnet does not reach out and

grab a distant nail across empty space. Instead, the magnet fills the space around it with a magnetic field, a real presence at every point, and the nail responds to the field right where the nail sits. The influence is not a spooky tug across a gap. It is carried, point by point, through a thing that genuinely fills the space in between.

Cast your mind back two chapters. Newton had described gravity perfectly but confessed he could not stomach the idea of one thing pulling another across pure emptiness with nothing in between. *I frame no hypotheses,* he had said, leaving the gap open. Faraday's field is the beginning of the answer Newton could not give. Space is not empty. It is filled with fields, invisible but real, and they are what carry the push and the pull from here to there. The bookbinder's apprentice, who knew no mathematics, had glimpsed the deepest furniture of the universe.

He just could not write it down. For that, the world needed someone who could see what Faraday saw *and* command the mathematics to capture it. That someone was a young Scotsman named James Clerk

Maxwell.

Maxwell Writes It Down

Maxwell did something that takes a rare kind of genius: he took Faraday's pictures, the invisible lines, the fields filling space, and the scattered laboratory discoveries of Oersted and Faraday and others, and he forged them into a single, complete, mathematical description of electricity and magnetism. Boiled down to their meaning, his laws make four plain statements about the world.

First: electric charges are the sources of electric fields. Positive and negative charges sit at the ends of electric field lines, the way taps and drains sit at the ends of flowing water. Electricity has sources you can point to.

Second: magnetism has no such sources. Cut a magnet in half and you do not get a separate north pole and south pole; you get two smaller magnets, each with its own north and south. No matter how finely you slice, you never find a lone magnetic pole. Magnetic field lines never start or stop. They only ever form closed loops.

Third, which is Faraday's discovery made exact: a changing magnetic field creates an electric field. Move the magnet, make electricity.

And fourth, which is Oersted's discovery made exact, plus a piece of pure genius that Maxwell added himself: an electric current creates a magnetic field, and so, Maxwell insisted, does a merely *changing* electric field, even with no wire and no current at all, even in perfectly empty space.

That fourth point, Maxwell's own addition, looks like a small tidy-up. It was the key that unlocked the universe.

And here is the part that ought to give you a chill about how this kind of discovery actually happens. Maxwell did not add that fourth piece because some experiment forced him to. No experiment had shown it. He added it because, without it, the equations quietly contradicted themselves: they broke the old, ironclad rule that electric charge can never simply appear or vanish. The mathematics had a gap, and Maxwell filled it with a term that logic demanded but that no one had ever seen, a changing electric field, all on its own, conjuring magnetism out of empty space. He put it in to make the equations honest with themselves. And that single invented term, added purely for the sake of consistency, was precisely the term that lets the fields leapfrog through space as a self-sustaining wave. He was driven to predict the very possibility of light by an act of careful bookkeeping, before he had any idea that was what he was doing. The mathematics knew about light before the man did.

Written in the compact modern form, the four laws are the equations at the head of this chapter, and you do not need to read the symbols to feel their power. The upside-down triangle is just shorthand for "how this field spreads out and curls through space," and the rest says, in the tightest possible code, exactly the four sentences you just read. Four lines. The complete behaviour of electricity and magnetism everywhere in the universe, for all time. But the real miracle was still hiding inside them, and Maxwell was about to

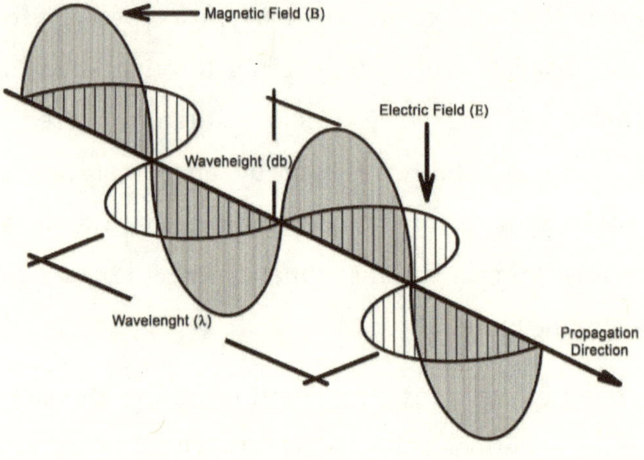

Figure 6: Figure 6.1 — An electromagnetic wave: the electric and magnetic fields oscillate perpendicular to each other and to the direction of travel.

find it.

The Ripple That Makes Itself

Look again at the third and fourth laws, and put them side by side, because together they do something extraordinary.

A changing magnetic field creates an electric field. A changing electric field creates a magnetic field.

Now let your mind run that loop forward. Suppose, somewhere, a magnetic field changes. By the third law, it creates an electric field. But that electric field is itself springing into being, which means it too is changing, and by the fourth law a changing electric field creates a magnetic field. Which is now also changing, so it creates another electric field, which creates another magnetic field, on and on, each one giving birth to the next, leapfrogging over and over.

You have a self-sustaining ripple. A disturbance in the electric and magnetic fields that builds its own road as it travels, each field forever regenerating the other, needing no wire, no medium, nothing to carry it. It can race through completely empty space, feeding on itself, forever. An electromagnetic wave.

And here Maxwell did the calculation that changed everything, and it is one you can follow, and even finish for yourself.

His equations did far more than say the wave exists. They pinned down exactly how fast it must travel, and left no choice in the matter. Each of the two laws makes a precise demand. Faraday's law fixes how much electric field a given changing magnetic field must produce. The other law fixes how much magnetic field a given changing electric field must produce. For a single ripple travelling through space, those two demands have to agree with each other at every instant, the field each law conjures must be exactly the field the other law needs. And it turns out they agree at one travelling speed and one only. Pick any other speed and the two laws flatly contradict one another, and the wave cannot hold together at all. So the speed of the ripple is not a free choice. It is forced.

And it is forced to be a value built from just two

numbers. One says how stiffly empty space resists an electric field; the other, how stiffly it resists a magnetic one. We write them ε_0 and μ_0, and here is the crucial thing: both had already been measured, on ordinary laboratory benches, with nothing more than charged metal plates, coils of wire, and magnets. Neither measurement had the faintest thing to do with light. Maxwell's equations then demand that his wave travel at exactly

$$c = \frac{1}{\sqrt{\mu_0 \varepsilon_0}}$$

Now do the thing this whole book keeps promising you, and check it with your own hand. The two bench-measured numbers, in the standard scientific units, are about

$$\varepsilon_0 \approx 8.85 \times 10^{-12} \qquad \mu_0 \approx 1.26 \times 10^{-6}$$

Multiply them together. Take the square root. Divide that into one. You need trust no one; any calculator will do it. Out comes a speed of very nearly

299,800,000 metres per second.

Two dry electrical measurements, taken with wires and magnets by people who were not thinking about light in the slightest, combined into a single number. And that number is not merely *close* to the speed of light. It *is* the speed of light, matched to every decimal place anyone has ever been able to measure. The proof is sitting in your calculator: take how space resists electricity, take how it resists magnetism, fold them together as the equation says, and you are handed the speed of a sunbeam.

(The airtight version of why those two laws agree at only one speed needs a little calculus, the kind that adds up how the fields change from point to point. But the result is exactly this, and the arithmetic that lands you on the speed of light is real, and yours to do.)

Sit with what had just happened. Maxwell had not been studying light. He had been studying sparks and coils and compass needles, electricity and magnetism. And his equations had quietly produced a wave that travelled at exactly the speed of light, built out of nothing but changing electric and magnetic fields. There was only one possible conclusion, and it was staggering. Light *is* this wave. Light is an electromagnetic disturbance, an interlocking ripple of electric and magnetic fields, propagating through space. The thing that lets you see, that pours from the Sun, that has flooded the world every dawn since the beginning, is electricity and magnetism, taking turns, racing along together.

A man with a pen had worked out what light is made of, by studying magnets and wires, and he was right. Humanity had wondered about light for as long as it had wondered about anything. Maxwell answered it with an equation.

The Whole Rainbow, and Far Beyond

There was more, and it was just as bold. If light is an electromagnetic wave, then its colour is simply how rapidly the fields wave up and down, its frequency.

Red light waves more slowly, violet light more quickly, and the colours of the rainbow are nothing but electromagnetic waves at slightly different rates, the way different notes are just air vibrating at different rates.

But then, Maxwell's equations placed no limit on that rate. There was no reason the waving had to stay within the narrow band our eyes happen to catch. There should be electromagnetic waves slower than red and faster than violet, waves no eye could ever see, but waves all the same, obeying the same four laws, travelling at the same speed.

He was right about that too. Below red sits infrared, the warmth you feel from a fire, and below that, microwaves and the radio waves that carry every broadcast. Above violet sits ultraviolet, the part of sunlight that burns your skin, and beyond it X-rays, which pass through flesh to photograph the bone, and gamma rays from the hearts of dying stars. All of it, from the radio mast to the dentist's X-ray, is the *same thing* as the light from your bedside lamp. One single phenomenon, stretched across an enormous range, almost all of it invisible to us, and almost all of it unsuspected until Maxwell's equations insisted it had to be there.

The proof came a few years after Maxwell's death. A young German physicist named Heinrich Hertz built apparatus to make electric sparks jump, and then detected invisible waves leaping across his laboratory from one device to another, waves with exactly the properties Maxwell's equations demanded. He had generated and caught radio waves, conjured straight out of the mathematics. Asked what use his discovery might have, Hertz reportedly shrugged that it had none whatsoever. Within a few decades it would carry the human voice around the planet.

One Force, Where There Were Two

Step back from the light and the signals for a moment, and feel the size of what Maxwell actually did, because it is bigger than any single discovery in this chapter.

He took two forces and proved they were one.

Before him, electricity and magnetism were two separate chapters in the book of nature, two different forces with two different sets of rules. After him, both were gone, replaced by a single thing: the electromagnetic field. Electricity and magnetism turned out not to be two forces at all, but two faces of one force, like a coin showing a head or a tail depending only on how you turn it. What looks like a purely electric effect to someone standing still can look partly magnetic to someone moving past, and the reverse. They are one substance, seen from different angles.

This is called unification, and it is one of the most powerful ideas in all of science. To unify is to discover that two things you believed were separate are secretly the same, that the world has fewer working parts than it seemed to, that beneath the clutter there is unity. We have felt this thrill already in this book. Newton unified the heavens and the Earth, the falling apple and the orbiting Moon, under one law of gravity. Maxwell's unification folded electricity, magnetism, and light together into a single field.

And it set the pattern for everything physics has attempted since. Once you have watched two forces collapse into one, you cannot stop wondering how many of the others are hiding the same secret. Could all the forces of nature be different faces of one deeper force? That question, lit by Maxwell's success, has driven physics for over a century. Decades later, electromagnetism was itself joined to one of the forces hiding inside the atom, the two revealed as a single "electroweak" force at high enough energies, an echo of Maxwell's triumph in a new key. The famous dream of a "theory of everything," one master equation from which all the forces pour, is nothing more than the dream of finishing the job Maxwell began: to keep unifying until only one thing is left.

So Maxwell did not merely tell us what light is. He showed all of physics what its deepest ambition could be. He proved that the universe can be simpler than it looks, that two can become one, and that the instrument for uncovering such hidden unity is exactly the kind of mathematics this whole book is about. Every chapter so far has murmured the same suspicion, that beneath the staggering variety of the world there lies a small, connected, comprehensible core. Maxwell did not murmur it. He proved a piece of it outright, and handed physics its marching orders for the next two hundred years.

The World It Built

Of all the equations in this book, this is the one whose fingerprints are most thickly smeared over the world you live in. It is not an exaggeration to say that Maxwell's four lines built the modern age.

Every electric motor and every generator runs on the third and fourth laws, magnets and changing fields making currents and forces; the entire electrical grid, every appliance you own, is Faraday and Maxwell made into metal. And then there is everything that flies invisibly through the air. Every radio and television broadcast. Every mobile phone call and text. Every wifi connection in every home and café on Earth. Radar that guides aircraft and tracks storms. The satellite signals that fix your position to within a few metres. Bluetooth, garage remotes, the tap of a card on a reader. All of it is engineered electromagnetic waves, deliberately shaped and sent and caught, every one of them a ripple obeying Maxwell's equations exactly. An antenna is nothing but a piece of metal arranged to make Maxwell's waves on purpose, and another arranged to feel them arrive.

It did not take long for someone to grasp what that was worth. Hertz had made his invisible waves leap a few metres across a laboratory bench and pronounced them useless. Within about fifteen years, a determined young Italian named Guglielmo Marconi had pushed those same waves clean across the Atlantic Ocean. In 1901, a signal sent from Cornwall, on the western tip of England, was caught on the far shore in Newfoundland, having crossed thousands of kilometres of open sea with no wire of any kind between the two. Experts had argued it was impossible, that the waves would fly off straight into space rather than follow the curve of the Earth. They made the crossing anyway. For the first time in human history, a message had travelled across an ocean at the speed of light with nothing physical joining the two coasts at all. The wireless age had begun, and every scrap of it stood on Maxwell's four lines and Hertz's sparks.

If you have ever worked with any of this, with the signals moving through a network, with the radio links and the spectrum and the careful art of getting a clean wave from one point to another, then you have spent your working life inside this chapter, whether or not Maxwell's name was ever mentioned. Every bit that crosses the air crosses it as an electromagnetic wave. The whole humming, connected, wireless world is four equations, switched on.

Is It Really True?

As true as anything in physics, and tested more ruthlessly than almost any idea humans have ever had. Maxwell's equations have been checked for over a century and a half, in every laboratory and every device, and the entire functioning of the electrical and electronic world is a continuous, planet-wide experiment confirming them every second. When your phone finds a signal, Maxwell passes the test again.

And, in the honest spirit of this whole book, they are not quite the final word, and the place they bend is one of the most beautiful in all of science. Maxwell's fields are smooth and continuous, like flowing water. But probe light closely enough, at the tiniest scales, and it stops behaving like a smooth wave and starts arriving in tiny, indivisible lumps, packets of energy we call photons. Light is somehow both the smooth wave Maxwell described and a hail of particles, depending on how you look at it. Reconciling the two required a deeper theory, quantum electrodynamics, which folds Maxwell's fields into the strange rules of the quantum world, and which has since become the most precisely tested theory in the history of science, agreeing with experiment to more decimal places than any other thing humans have ever predicted.

But notice: the quantum theory did not throw Maxwell away. It contains him. For everything larger than an atom, for every antenna and every sunbeam, Maxwell's four equations are exactly, gloriously right. They simply turned out to be the smooth, large-scale face of something even stranger underneath, and we will go down into that stranger world soon enough.

The Quiet Polymath

It is worth meeting James Clerk Maxwell properly, because he may be the greatest physicist most people have never heard of, and because the man behind these four equations was extraordinary in ways that had nothing to do with electricity.

He was born in Edinburgh in 1831, into a comfortable Scottish family, and was curious to the point of comedy from the very start. As a small child, his constant demand about anything that caught his eye was "what's the go of it?", and if a first answer did not satisfy him, "but what's the *particular* go of it?" He wanted the real working, the actual mechanism, of every single thing in the world. He never lost it.

His mind flatly refused to specialise. While still in his twenties, asked to explain how the rings of Saturn could possibly be stable, he worked out, with nothing but mathematics, that they could not be solid and could not be liquid, and so must be made of countless separate chunks, each a tiny moon orbiting on its own. He was right. Nobody could confirm it until spacecraft flew past Saturn more than a century later and found exactly the swarm of orbiting rubble he had described from his desk.

He all but founded the science of colour vision, working out that the eye senses only three primary colours and builds every other shade from mixtures of them. And in 1861, to prove it, he made the first colour photograph in history, an image of a tartan ribbon, by combining three black-and-white photographs taken through red, green, and blue filters. Every colour screen you have ever stared at, every photo on your phone, every television, still works on the principle Maxwell demonstrated with that ribbon.

And then, in a feat that would on its own have earned him a place among the immortals, he helped build an entirely separate branch of physics: the statistical understanding of heat. He saw that the temperature and pressure of a gas are just the blurred-together effect of countless molecules tearing about at a whole spread of different speeds, and he worked out the mathematics of how those speeds are shared out. We will meet that idea again near the end of this book, when we come to heat and the strange one-way arrow of time. The same quiet man who told us what light is also told us what heat is.

He did all of it without fuss or self-promotion, by every account kind, modest, and deeply decent, devoted to his wife, fond of terrible puns and comic verse, endlessly generous to younger scientists. And he did it in a tragically short life. He died of stomach cancer in 1879, the same illness that had taken his mother when he was a boy, at just forty-eight. Einstein, who built his own revolution directly upon Maxwell's electromagnetism, kept a portrait of him in his study beside those of Newton and Faraday, and called the shift in our picture of reality that Maxwell set in motion the most profound and fruitful that physics had known since Newton himself.

He is not a household name. He should be. Ask most people to name a great scientist and they will reach for Newton, or Einstein, and never think of the quiet Scot who sat between them and worked out what light is, what colour is, and what heat is, and what the rings of Saturn are made of, before dying younger than many people's parents.

The Name on the Door

We call them Maxwell's equations, and unlike some names in this book, this one is richly deserved. But, as always, the door has more than one name behind it.

The first belongs to Faraday, the unschooled bookbinder's apprentice who could not do the mathematics but who *saw* the fields with his mind's eye and handed Maxwell the central idea. It is one of the loveliest partnerships in science: the working-class experimental genius who could picture the invisible but not write the equations, and the gentleman mathematician who could write the equations but needed the picture first. Neither could have done it alone. Maxwell, to his great credit, said so plainly, and revered the older man.

And then there is the curious fact about the famous

four equations themselves. Maxwell did not write them in the tidy, four-line form printed at the start of this chapter. His own version, set down in the 1860s and 70s, sprawled across some twenty equations in twenty quantities, a dense thicket that even most physicists of the day found forbidding, and it sat there, admired but largely unused, for years. It was only in 1884, five years after Maxwell had died young, at just forty-eight, that a brilliant, reclusive, self-taught English engineer named Oliver Heaviside took that sprawling mass and, using the tools of vector mathematics, crushed it down into the four clean, symmetric lines the world now knows by heart. The equations we call Maxwell's were given their famous shape by another man's hands, after their author was in his grave.

So here, one more time, is the pattern this book keeps uncovering. The deepest truth in the chapter, that light is an electromagnetic wave, belongs unmistakably to Maxwell. But the apple-and-Newton tangle of names clings even here: a self-taught visionary who supplied the idea, a young Scot who found the truth, an eccentric engineer who carved the final famous form, and a German who proved it all real only after the discoverer had died. The names jostle and overlap and quarrel down the years. The four equations do not care. Light has been an electromagnetic wave since the first star lit, indifferent to whose name we eventually wrote above it.

So Look

You do not need to take any of this on faith. You are soaking in the proof.

The warmth you feel on your face in sunlight is electromagnetic waves, made in the Sun, striking your skin and shaking it, after an eight-minute journey across empty space at the only speed these waves know. The colours of everything you can see are your eyes sorting those waves by how fast they ripple. Every time a message lands on your phone with no wire attached, an electromagnetic wave that was a radio signal moments ago has just arrived and been caught by a scrap of metal inside, exactly as Maxwell's equations promised and Hertz first proved. You could lay a compass beside a wire and watch electricity nudge the needle, the way Oersted did two centuries ago, and see the two great mysteries reach across and touch with your own eyes.

You live, every second, bathed in the thing this chapter describes. Light to see by, warmth to feel, and an invisible weather of signals carrying the voices of the whole human race straight through your body. It is all one thing, and four equations caught it.

But leave this chapter holding one loose thread, because it is about to unravel the universe. Buried inside Maxwell's equations is that speed, the speed of his waves, the speed of light. And the equations are strangely stubborn about it. They do not say light travels at that speed *relative to anything*. They simply say it travels at that speed, full stop, as a fact built into the fabric of space itself. Which raises a question so innocent it sounds like a child's, and so dangerous it would shatter our oldest ideas of space and time. If you chased a beam of light, racing after it as fast as you could, how fast would it then seem to flee from you?

The obvious answer is wrong. And a young clerk in a patent office, turning that exact question over in his mind, was about to follow it all the way down, until time itself began to bend.

Chapter Seven

Mass-Energy Equivalence

The Sleeping Giant

$$E = mc^2$$

There is one equation in this book you already knew before you opened it. You have seen it on T-shirts and film posters, in cartoons and advertisements, chalked on blackboards in the background of a hundred films to signal, wordlessly, *clever*. Three letters, one small number, and the most famous name in the history of thought. $E = mc^2$.

It may be the most recognised string of symbols our species has ever produced. And it is also, almost certainly, the one the fewest people could actually explain. We know its face the way we know a celebrity's face, without ever having met the person behind it. So let us finally meet it properly.

Here is what it says, and it is not a gentle thing to be told. The coin in your pocket, the last mouthful of coffee in your cup, the ring on a finger, each of these holds, locked away inside it, enough energy to level a city. Not as a figure of speech. In hard, countable joules. Ordinary matter, the dull stuff the whole world is built from, is not dull at all. It is energy, packed unimaginably tight and held unimaginably still, a fire frozen so hard that it looks like a stone. This equation is the key to the lock. It tells you exactly how much fire is sleeping in the stone. And the answer is far, far more than anyone had ever dreamed.

The Thread From the Last Chapter

We arrive here on a thread I deliberately left dangling. At the close of the last chapter, Maxwell's equations handed us a strange and stubborn fact: light travels at one fixed speed, and that speed is built into the laws of nature themselves, not measured against anything else.

A teenager named Albert Einstein became obsessed with a simple question about it. What would happen, he wondered, if you chased a beam of light? Run faster and faster, until you were flying along beside it. Surely then the light would seem to slow, to hang frozen in the air next to you, like a wave you had finally caught up with.

But Maxwell's equations flatly refused to allow it. They insisted the light must always race away at its full speed, no matter how hard you ran. You can never catch it. You can never even gain on it. Sprint toward an oncoming beam, or flee from one, and it closes on you or leaves you at exactly the same speed every time, as though your own running counted for nothing.

That is impossible by every ordinary instinct we have about motion. Einstein's genius was to take it seriously anyway, to assume the impossible-sounding fact was simply *true,* and then follow it without flinching wherever it led. Where it led was the quiet demolition of common sense. If the speed of light is the same for everyone, then space and time themselves must bend and stretch to keep it so. Moving clocks must run slow. Moving rulers must shrink. Two people rushing past one another cannot even agree on whether two distant events happened at the same moment. This was his theory of special relativity, published in 1905, and it rebuilt the stage of space and time from the boards up.

And then, almost as an afterthought, in a short note added a few weeks later, Einstein noticed that his new theory carried a consequence he had not gone looking for. If space and time behaved like this, then mass and energy could not be two separate things. They had to be one and the same.

Reading the Three Letters

Let us take the equation apart, gently, the way we have taken apart every other one.

E is energy: the capacity to do things, to move, heat, push, light, break. m is mass, the amount of stuff in a thing, the very stubbornness we met in Newton's chapter, the quantity that fights being shoved. And c is the speed of light, that fixed cosmic speed limit Maxwell uncovered, about three hundred million metres every second.

The whole equation reads: the energy hidden inside something equals its mass, multiplied by the speed of light, multiplied by the speed of light again. Energy equals mass times c squared.

The whole shock lives in that little c^2. The speed of light is already an enormous number. Square it, multiply it by itself, and you get a monster:

$$c^2 \approx 90{,}000{,}000{,}000{,}000{,}000$$

That number is the exchange rate between mass and energy, and it is a brutal one. It means every single kilogram of mass is worth that monstrous pile of joules. A speck of matter buys a mountain of energy. Mass is simply energy with an astronomical price tag, energy that has been concentrated and frozen so completely that a thumbnail of it can hold the violence of a bomb. The equation is not really saying that mass and energy are *related*. It is saying they are the *same thing*, two forms of one single substance, and c^2 is just the number that converts between the units, the way a fixed rate converts dollars into cents. It is only that this particular exchange rate is the most lopsided in the universe.

The Proof in a Box

That is a staggering claim, that a stone is frozen fire. This is a book about showing the working, so I will not ask you to take it on trust. I am going to derive $E = mc^2$ from scratch, using only two things we have already built in earlier chapters. Follow closely. It is one of the most beautiful arguments in all of physics, and Einstein himself gave a version of it.

Picture a box. A sealed box, floating alone and perfectly still in deep space, far from everything, with nothing pushing or pulling on it. Let it have mass M and length L.

Now the left wall of the box flashes a single pulse of

light across to the right wall, which absorbs it. That is the entire experiment. One flash, crossing the box from left to right. Watch what it forces to happen.

We need two facts, and we own both already.

The first is from Maxwell. Light carries a push. A pulse of light, though it has no weight you could feel in your hand, carries momentum, and when it is fired off it shoves back on whatever sent it, exactly as a bullet kicks back on the gun. This is real and measurable. It is why a comet's tail always streams away from the Sun, blown back by the faint pressure of sunlight, and why engineers can build solar sails that coast through space on nothing but starlight. The momentum carried by a pulse of energy E is that energy divided by the speed of light, $p = E/c$.

The second fact is from Newton. Momentum is conserved, and so the balance point of a lonely system, its centre of mass, cannot move on its own. Nothing outside the box is touching it. So whatever goes on inside, the box's centre of mass must stay exactly where it began. It is simply not allowed to drift.

Now let us follow the flash, step by step.

The instant the left wall fires the light rightward, the box must recoil leftward to keep the total momentum at zero, just as you rolled backwards on the skateboard when you threw the ball. The box drifts left at a tiny speed equal to the light's momentum shared over the box's mass:

$$v = \frac{p}{M} = \frac{E}{Mc}$$

The light takes a little time to cross the box, its length divided by its speed:

$$t = \frac{L}{c}$$

And during that little time, the slowly drifting box slides to the left by a small distance, its speed times that time:

$$\Delta x = v \times t = \frac{E}{Mc} \times \frac{L}{c} = \frac{EL}{Mc^2}$$

Then the light strikes the far wall, is soaked up, and

the box stops dead. The flash is over. And here is the problem that cracks the whole of old physics wide open. The entire box has shifted to the left. But nothing outside ever touched it, and we agreed the centre of mass of a lonely thing is forbidden to move. Something must have travelled to the right to make up for the box's drift, to keep the balance point pinned exactly where it was.

The only thing that went to the right was the light.

So the light must have carried mass with it, from the left wall to the right wall. Real mass, just enough to keep the books balanced. Picture the centre of mass as the pivot of a seesaw. The box, with its large mass M, slid left by the small distance Δx. For the pivot to stay put, the light, carrying some small mass m, must have travelled right across the box's full length L, so that the two exactly cancel:

$$M \times \Delta x = m \times L$$

Now simply feed in the distance the box moved, the Δx we worked out a moment ago:

$$M \times \frac{EL}{Mc^2} = m \times L$$

The M on the left cancels. So does the L on each side. And what is left standing, when the dust clears, is this:

$$m = \frac{E}{c^2}$$

The pulse of light, carrying energy E, also carried mass equal to E/c^2. Energy *has* mass. And turning that statement around gives us the famous one:

$$E = mc^2$$

There it is. The most famous equation in the world, built with your own hands out of a box, a flash of light, and two laws you already trusted. No magic anywhere. Every step open to inspection. Light carries a push, the balance point cannot move, and the only way to keep the books straight is for energy itself to weigh something. From that, and nothing more, the giant wakes.

Be the Judge

Now take the equation out for a drive, and feel exactly why it frightened the people who first understood it.

Pick up something small. A paperclip weighs roughly a gram, which is a thousandth of a kilogram. Let us ask the equation how much energy is frozen inside it. Multiply the mass by c^2:

$$E = mc^2 = 0.001 \times 90{,}000{,}000{,}000{,}000{,}000$$

$$E = 90{,}000{,}000{,}000{,}000 \text{ joules.}$$

You can do that on any calculator in two seconds: nought point nought nought one, times nine with its sixteen zeros. Ninety trillion joules, sitting quietly inside a paperclip.

How much is that? It is roughly the energy of twenty thousand tonnes of TNT. It is, near enough, the bomb that flattened a city. Inside one paperclip. The only reason the paperclip does not casually explode is the single mercy in the whole affair: there is no easy way to unlock that energy. Matter holds its fire ferociously tight, and only a few violent processes, the ones inside bombs, reactors, and stars, can prise loose even a sliver of it. The energy is utterly real. The lock is just very, very good.

Run the same sum on yourself, if you have the stomach for it. A person of seventy kilograms is, by this equation, worth more than a billion billion joules, the equal of a hundred thousand of those city-flattening bombs. You are, every one of us is, a sleeping giant wrapped in skin, kept harmless only because our atoms flatly refuse to give the energy up. We walk around carrying the power of stars, and we feel nothing at all, because the door stays shut.

Why the Sun Burns

I made you a promise, at the very start of this book, that one of these equations finally explains why the Sun shines and does not go out. This is that equation, and this is that moment.

For most of history the Sun should have kept thoughtful people awake at night. It pours out a ceaseless torrent of light and heat, and has done so, we now know, for more than four billion years, long enough for life to crawl out of the sea and eventually build telescopes to look back at it. What could possibly burn for that long? A Sun made of coal and set alight would burn itself to ash in a few thousand years. An ordinary fire was hopelessly, absurdly too feeble. The Sun's furnace had to run on something nobody had ever seen.

$E = mc^2$ is the answer. Deep in the Sun's crushing core, hydrogen, the lightest and simplest stuff there is, is squeezed so violently that its nuclei fuse together into helium. And here is the trick the equation lays bare: when the bits of hydrogen merge into helium, the helium that comes out weighs a tiny bit *less* than the hydrogen that went in. A little mass has simply gone missing. It has not vanished into nothing. It has been converted, at that monstrous exchange rate, into pure energy, and flung outward as sunlight.

The mass lost in each single merging is laughably small. But the Sun is vast beyond imagining, and it does this an unthinkable number of times every second. Add it all up, and the Sun is turning about four million tonnes of itself into pure energy every single second. Four million tonnes of the Sun's own substance, every tick of every clock, becoming the light that warms your face, grows the wheat, lifts the weather, and made every living thing that has ever lived possible. The Sun is not burning in any ordinary sense. It is slowly, steadily converting its own mass into light, and it carries so vast a store of mass to spend that it will go on doing so for billions of years yet.

Every sunrise you have ever watched was $E = mc^2$, written in fire across the sky. The warmth on your skin on a clear day is a few grams of the Sun, which left it eight minutes ago as energy, arriving at last. We owe our entire existence to a slow and gentle leak of mass into light, a hundred and fifty million kilometres away.

The Same Fire, Unleashed

There is a darker face to all of this, and the book would be dishonest to look away from it.

The very same equation that lights the Sun and feeds all of life is also the one that taught humanity how to destroy itself. If a sliver of mass can become a flood of energy, then learning to unlock that mass on purpose, here on Earth, means power beyond anything in history, for good and for ruin alike.

The gentle face is the nuclear power station. Split heavy atoms apart, instead of fusing light ones together, and again a little mass goes missing and reappears as heat. A single kilogram of nuclear fuel can release as much energy as thousands of tonnes of coal. The lights of whole nations are kept on by the steady vanishing of a few grams of mass.

The terrible face arrived first. In 1945, the energy that destroyed the city of Hiroshima came from the conversion of less than a single gram of matter, under a gram, lighter than a paperclip, turned wholesale into energy in a fraction of a second. That is the equation's verdict laid utterly bare. The most beautiful and concise statement in all of science, the secret of the stars, is also the cold arithmetic of the worst weapon ever built.

Einstein's own part in this is one of history's heavy ironies, and it is worth getting right, because it is so often told wrong. He did not build the bomb. He was not part of the project that made it; he was thought a security risk and kept well clear of it. What he did was sign a letter, in 1939, warning the American president that such a weapon might be possible and that Germany might already be chasing it. He spent much of the rest of his life dismayed at what came after, and called the signing of that letter the one great mistake of his life. The man who found the deepest unity in nature watched it turned into a means of mass death, and grieved. The equation, of course, did neither good nor evil. It only ever stated a fact. What we chose to do with the fact was always going to be on us.

You Are Mostly Frozen Energy

Step back from bombs and stars for a moment, because the equation has something quietly astonishing to say about ordinary things too, and about you.

If mass and energy are truly the same, then anything that gains energy must gain mass, and anything that loses energy must lose mass. This is not a poetic flourish. It is literal. A hot cup of coffee weighs the very tiniest sliver more than the same cup once it has cooled. A wound-up spring weighs more than a slack one. A charged phone battery is a shade heavier than a flat one. The amounts are so absurdly small, divided as they are by that vast c^2, that no kitchen scale could ever hope to show them. But they are real, and they have been measured in the cases where the energies are large enough to see. Energy has weight. Always.

And now the strangest turn of all. You might assume that the mass of an object is just the sum of the masses of the little pieces it is built from. It is not. Take a single proton, one of the particles sitting in the heart of every atom in your body. It is made of smaller things called quarks. And here is the shock: those quarks, added together, account for only about one percent of the proton's mass. The other ninety-nine percent is not made of *stuff* at all. It is the *energy* of the ferocious force binding the quarks together, the violent churning field caged inside the proton, registering as mass through this very equation.

Let that settle. Nearly all the mass of every proton in you, and therefore very nearly all of *your* mass, the figure you read on the bathroom scale, is not solid matter in any honest sense. It is bound-up energy, frozen by $E = mc^2$ into the convincing appearance of solid weight. You are not a thing made of stuff with a little energy tucked inside. You are very nearly pure energy, knotted up tightly enough to stand up and read a book. The stone really was frozen fire all along. And so, it turns out, are you.

Is It Really True?

It may be the most thoroughly confirmed equation ever written, because we have managed to check it from both directions at once.

We have watched mass turn into energy, in every nuclear reactor humming on the grid, in every star scattered across the sky, and, terribly, in the deserts and the two cities where the bombs were tested and used. The numbers match the equation every time, to as many decimal places as we can measure.

And we have watched the equation run *backwards*, energy turning into mass, which is somehow even more astonishing. In the great particle accelerators, the giant rings such as the one buried beneath the border of France and Switzerland, physicists take tiny particles, pump them full of energy until they are moving a whisker below the speed of light, and then smash them head on. Out of the collision, out of the raw energy alone, brand new particles spring into being, heavier than anything that went in. Mass, conjured out of pure energy, made to order, in precisely the amount $E = mc^2$ says you should get for the energy you spent. It happens millions of times a day. Energy congeals into matter in front of our instruments, on demand. The equation is not just a description of distant stars. It is the daily working tool of every laboratory trying to learn what the world is made of.

The Name on the Door

The name on this door is the most famous in the history of science, and for once the fame is fully earned. But even here, the door has more written on it than the legend lets on.

Albert Einstein arrived at all of this in 1905, in what is now simply called his miracle year, and the circumstances are almost comic. He was not a professor. He had failed to land an academic post of any kind, and was working as a junior clerk in the patent office in Bern, examining other people's inventions by day and quietly rebuilding physics in his spare hours. In that single year, in those spare hours, this overlooked clerk published the work that explained the photoelectric effect, that proved atoms are real, that overthrew our oldest notions of space and time, and that, in a four-page afterthought, announced that mass and energy are one. It remains the greatest year of thinking any single person has ever had.

And since this is a book about checking the stories we are handed, let us puncture two of them.

The first is the comforting legend, repeated to struggling students everywhere, that Einstein failed mathematics at school. It is false. He was superb at mathematics from boyhood, and had taught himself calculus, for pleasure, by his mid-teens. The myth seems to have grown out of a misread school report, and it survives because we badly want to believe that genius can hide behind a bad grade. It cannot, at least not here.

The second is subtler, and it concerns the equation itself. The clean, perfect $E = mc^2$ that the whole world can write from memory does not actually appear, in that form, anywhere in Einstein's 1905 paper. What he wrote was quieter and more careful. His paper carried the modest, almost timid title *Does the Inertia of a Body Depend on Its Energy Content?*, and it concluded that if a body gives off a quantity of energy, its mass falls by that energy divided by c^2. From this he drew the line that truly mattered: that the mass of a body is a measure of how much energy it holds. The tidy three-symbol formula we all carry in our heads is a later distillation of that idea, polished smooth by other hands over the years that followed. Even the most famous equation in the world was not, in the end, written in its famous form by the hand we give the credit to. And it had quiet forerunners besides. Others, the French mathematician Henri Poincaré among them, had glimpsed fragments of the link between mass and energy before him. Einstein's towering achievement was not to spot a fragment, but to grasp the whole, and to state it plainly and without exception: all energy has mass, all mass is energy, everywhere in the universe, no matter what.

So Feel It

You do not need a laboratory to meet this equation. You meet it constantly.

Every time the Sun warms your face, you are feeling mass turned into light, a few grams of a distant star arriving as energy. Every plant you have ever eaten was assembled by that arriving sunlight, which means every meal you have ever had was, at its root, $E = mc^2$ feeding you. And the next time you step on a scale and read the number, remember what that number mostly is: not stuff, but energy, bound up tightly enough to hold the shape of a person. You can run the sum yourself, the very one we did together, and know that the figure under your feet is very nearly a measure of the fire frozen inside you.

The giant in the stone never wakes for us, and we should be grateful for that. But it is in there. The equation tells us so, the Sun proves it every morning, and the calculators in our pockets let us check the size of the sleeping thing whenever we please.

Relativity had tamed the very fast and the very large. It told us how space and time bend, where the light of the stars comes from, and what matter most deeply is. It was a triumph of the grand scale, the cosmos seen whole and clear.

But all the while, another door had been quietly creaking open, leading the opposite way, *downward,* into the unimaginably small. Inside the atom itself, a far stranger revolution was waiting, one that would make even relativity look cosy and familiar by comparison. Down there, at the scale of the very particles we have just been weighing, things would stop having definite positions at all. A single particle would be here and there at once, its whereabouts nothing but a haze of maybe. And the next equation we meet does not tell us where anything *is*. It only, ever, tells us where it might be.

Chapter Eight

Schrödinger's Equation

The Cloud of Maybe

$$i\hbar\frac{\partial\Psi}{\partial t} = \hat{H}\Psi$$

Here is a fact that should not let you sleep. By the physics of the last seven chapters, you should not exist. Nor should this book, the chair, the air, or any solid thing at all.

The reasoning is short and brutal. An atom is a tiny, dense nucleus with electrons whirling around it, like planets around a sun. But an electron is an electric charge, and Maxwell taught us that an electric charge whipping around in a circle must pour out energy as light. Losing energy, the electron should spiral inward, faster and tighter, and crash into the nucleus. The sums are merciless: every atom in the universe should collapse in about a hundredth of a billionth of a second. All matter should have flickered out of existence the instant it formed.

It did not. You are still here. Atoms are rock solid, patient, eternal. Something keeps the electron from falling, something that the magnificent machinery of Newton, Maxwell, and Einstein could not name. To find it, physics had to go down, past the atom's edge, into the smallest rooms of reality, and there it found a world so strange that the people who discovered it spent the rest of their lives arguing about what they had seen. The law of that world is one short, fearsome-looking equation. It works to a precision matched by nothing else in science. And it describes a reality in which a thing can have no definite place at all.

The Clue in the Colours

The first clue had been sitting in plain sight for decades, in light.

Take a tube of pure hydrogen gas and run electricity through it, the way a neon sign works, and it glows. But pass that glow through a prism and you do not get a smooth rainbow. You get a few sharp, separate lines of pure colour, with darkness between them, always at exactly the same places. Every element does this, and every element's pattern is different and fixed. Sodium gives the orange of a street lamp. Neon gives its red. Each gas has a barcode of light, unique and unchanging, the same in a laboratory in Melbourne as in a star halfway across the galaxy.

This was a riddle. Light, as Maxwell showed, can have any colour at all, any frequency on a smooth and endless dial. So why would an atom only ever emit a few exact colours, and never the ones in between? It was as if the atom were a musical instrument that could play only certain notes, and was physically forbidden from sounding anything off the scale.

That word, *only*, is the whole mystery of this chapter. The atom can only hold certain amounts of energy, never the values in between. Its energy comes in steps, like rungs on a ladder, not in a smooth ramp. When an electron drops from a higher rung to a lower one, the atom spits out the leftover energy as a single particle of light of one exact colour. The sharp lines are the gaps between the rungs, made visible. But nobody could say *why* the rungs were there. Why should nature, which seemed so smooth everywhere else, suddenly come in lumps inside the atom?

The Reluctant Revolutionary

The atom was not the first place the lumps showed up. The very idea that energy comes in steps, rather than as a smooth flow, had been forced on a deeply unwilling physicist a quarter of a century earlier, and he hated it to his bones.

The problem he was wrestling with sounds almost too ordinary to matter: the colour of hot things. Everything glows when it is heated enough. A poker pulled from the fire glows dull red, then orange, then, hotter still, towards white. The Sun glows yellow-white, a cooler star red, a hotter one blue. The colour of the glow depends only on the temperature, in a precise and universal way, and by the end of the 1800s physicists badly wanted a formula for it.

When they tried to build one, using the trusted physics of the day, Maxwell's waves and the laws of heat, they got an answer that was not merely wrong but insane. The theory predicted that any warm object should blaze out an *infinite* amount of energy at the short-wavelength, ultraviolet end of the spectrum. By that reasoning, the gentle warmth coming off a kettle should be flooding the kitchen with a lethal torrent of ultraviolet and X-rays. The failure was so spectacular it earned a name: the ultraviolet catastrophe. Reality, of course, did no such thing. A warm kettle is perfectly safe to stand beside. The cherished old physics had collapsed completely on the simplest question imaginable, and nobody could see why.

In 1900, Max Planck found the fix, and it appalled him. He discovered that he could reproduce the real glow of hot objects exactly, matching every measurement ever made, but only by swallowing one outrageous assumption. He had to suppose that energy is not handed out in a smooth, continuous stream at all, but only in tiny separate packets, lumps, each carrying an amount of energy set by its frequency:

$$E = hf$$

Here f is the frequency of the light, and h is a new constant of nature, fantastically small, that no one had ever needed before. It is now called Planck's constant, and it is the most important number in this chapter, the very grain size of reality. With energy forced to come in these lumps, the ultraviolet catastrophe simply melts away. The short-wavelength packets cost so much energy each that a merely warm object can almost never afford to make even one. The infinity vanishes. The formula fits the world perfectly.

Planck had blundered into the deepest secret of the small: that energy itself is grainy, that nature counts in whole coins rather than pouring a smooth liquid. The name for one of those coins, borrowed from the Latin for "how much," was *quantum*. The entire revolution that fills this chapter takes its name from Planck's reluctant little lumps.

And here is the part that belongs in this book above all. Planck did not believe it. He was a careful, conservative man, and he regarded his own assumption as a grubby mathematical trick, an act of desperation, as he later called it, to be quietly tidied away the moment someone found the proper smooth explanation hiding underneath. He spent years trying to undo his own discovery, to dissolve the lumps back into a continuous flow and rescue the old physics. He never managed it, because there was nothing to rescue. The lumps were real. It took the young Einstein, in that same miracle year of 1905, to take Planck at his word and declare that light itself genuinely travels in these packets, the photons we have already met. The man who lit the fuse of the quantum age spent much of his life trying to stamp it back out. He failed. Reality had the final vote, and reality counts in lumps.

The Outrageous Idea

The answer came from a young French aristocrat, Louis de Broglie, in a doctoral thesis so audacious his examiners were not sure whether to pass it.

He started from the trick we met two chapters ago. Light had always been a wave, and then it turned out to also arrive in particles, the little packets called photons. Light was somehow both. De Broglie asked the question nobody had dared to ask backwards. If a wave like light can behave as a particle, then can a particle, like an electron, behave as a wave?

It sounds like nonsense. We do not see cricket balls rippling. But de Broglie wrote down exactly how long the wave of any moving object should be. Its wavelength is Planck's constant, a fantastically tiny number written h, divided by the object's momentum:

$$\lambda = \frac{h}{p}$$

Look at what that little equation does. For anything large, a person, a ball, a speck of dust, the momentum p is enormous compared to the absurdly tiny h, so the wavelength comes out so unimaginably small that the waviness is utterly invisible. That is why we never see people ripple. But for an electron, which has almost no mass and so very little momentum, the wavelength works out to roughly the size of an atom. At that scale, the waviness is not a rounding error. It is everything.

And the moment you picture the electron as a wave wrapped around the nucleus, the rungs of the ladder explain themselves. I will come to exactly how in a moment, because it is the heart of this chapter and you can derive it yourself. But here was the seed: matter is wavy, and the wave is what had been missing.

What Is Doing the Waving?

De Broglie said matter was a wave. But a wave of *what*? A water wave is waving water. A sound wave is waving air. What is waving when an electron moves?

This is where Erwin Schrödinger came in. He took de Broglie's loose, dreamlike idea and built it into a proper machine. The story is almost too neat: giving a seminar on de Broglie's waves, Schrödinger was needled by a senior colleague, who remarked that if you are going to talk about waves like a grown-up, you had better have a wave equation for them. Weeks later, Schrödinger reportedly opened his next talk by saying: my colleague wanted a wave equation, well, I have found one. He had, and it now bears his name.

The wave he wrote down is denoted by the Greek letter psi, Ψ, and it is the strangest object in this book. It is not a wave in any stuff. It does not ripple through water or air or any material at all. The thing Ψ describes, in the interpretation that became standard, is *possibility itself.*

Here is the meaning, fixed shortly afterward by Max Born, and it is worth reading twice. The wave Ψ is spread out across space. Where the wave is tall, you are very likely to find the particle if you look. Where the wave is small, you are unlikely to. Where it is zero, you will never find it. More precisely, the height of the wave squared, at any point, gives the exact probability of finding the particle there.

And the word *probability* is doing something far heavier than it looks. It does not mean the electron is really at one definite spot and we are merely ignorant of which. It means the electron has *no* definite spot. It is genuinely smeared across all its possible positions at once, a cloud of maybe, until the instant something measures it. Only then does the cloud collapse to a single point, snapping to one location, with the odds set by the wave. Before you look, asking exactly where the electron is has no more meaning than asking what is north of the North Pole. There is no answer because there is no fact.

That is the world inside the atom. Not little planets on little orbits, but standing clouds of probability, humming around the nucleus.

The Two Slits

If a single experiment could hold the whole strange soul of this chapter, it would be this one, and it is simple enough to picture completely. Richard Feynman, one of the great physicists of the last century, said it contained the only real mystery of quantum mechanics, that everything odd about the quantum world is, in the end, this one experiment wearing a disguise.

Take a wall with two narrow parallel slits cut in it, side by side. Behind the wall, a screen that records wherever something strikes it. In front of the wall, a source firing tiny particles at it, one kind, all identical, say electrons.

Think first about what ordinary little balls would do. Fire a spray of pellets at the two slits. Some pass through the left slit, some through the right, and behind the wall you build up two simple stripes on the screen, one behind each slit. Two slits, two bands. Obvious.

That is not what electrons do.

When you fire electrons at the two slits, the screen lights up not with two bands but with many: a pattern of bright and dark bars marching across it, bright where electrons pile up, dark where almost none ever land, repeating over and over. This is the unmistakable fingerprint of *waves*. When two water waves spread out and overlap, in some places two crests meet and add into a taller wave, and in others a crest meets a trough and they cancel into flat calm. Light does it, sound does it, and here, impossibly, electrons are doing it too. Each electron is crossing as a spread-out wave, passing through *both* slits at once, the two halves overlapping and interfering on the way to the screen, heaping up where they reinforce and staying clear where they cancel. It is the cloud of maybe, printed as a pattern you can photograph.

Now comes the part that has robbed people of sleep for a hundred years.

Turn the source down. Down and down, until it is releasing electrons so slowly that only one is ever inside the apparatus at a time, each crossing completely alone before the next is fired. There is now no other electron anywhere for it to interfere *with*. And sure enough, each one arrives at the screen as exactly what it is, a single particle, leaving one tiny dot. One electron, one dot. Fire a single electron and you learn nothing. But fire them for hours, one lonely electron at a time, and watch the dots slowly gather.

The stripes come back.

Dot by patient dot, the same striped interference pattern assembles itself out of thousands of separate, isolated landings. Which means each electron, travelling utterly alone with nothing else in the box, somehow went through *both slits at once* and interfered *with itself*. A single particle, on its own, spread through both openings as a wave of maybe, the two halves of its own cloud overlapping, and that overlap decided where its one dot was allowed to fall. It is not that half the electrons sneak left and half sneak right. Every last one goes through both, as a wave, and then arrives as a particle. There is no picture of this in ordinary life. There is no marble that does it. There is only the equation, and the equation is right.

And then the experiment lands its final, chilling blow. Suppose you decide to end the madness by simply *watching*. You place a tiny detector at the slits to catch which one each electron truly goes through, to settle once and for all this nonsense about both at once.

The instant you do, the stripes vanish.

With a watcher at the slits, every electron is caught going cleanly through one slit, left or right, like a well-behaved particle, and the screen builds up just two plain bands, exactly as dull little balls would make. The interference, the waviness, the going-through-both, all of it disappears the moment you look. Switch the detector off, and the stripes flood straight back. The electron behaves as a spread-out wave of possibility right up until the act of measurement pins it down, and at that instant the cloud of maybe collapses to a single hard fact: this slit, not that one, here, not there. Knowing *which* destroys the wave. The asking changes the answer.

That is the cloud of maybe, caught red-handed. The electron has no single path until something forces one upon it. Between leaving the source and being seen, it is not in one place taking one route; it is the whole spread of its possibilities at once, precisely the wave Ψ that we now need a law to govern. So what is the rule that steers it?

Reading the Equation

So what is Schrödinger's equation itself? It is the rule that tells the cloud what to do.

Think back to Newton. $F = ma$ was the law of motion for ordinary things: tell me the forces on a thing now, and the equation tells you how it moves next. Schrödinger's equation is the $F = ma$ of the quantum world, but for the wave of possibility. Tell it the shape of Ψ now, and it tells you the shape of Ψ a moment later. It is the law that steers the cloud of maybe through time.

$$i\hbar\frac{\partial\Psi}{\partial t} = \hat{H}\Psi$$

You are not meant to read that the way you read a sentence, and you do not need to. The left side, in plain words, is "how the wave is changing from one instant to the next." The right side is a compact instruction that says "and here is what its own shape and surroundings force that change to be." The $\hbar$, "h-bar," is just Planck's constant again, the fundamental grain size of the quantum world, the number that sets how lumpy reality is down here. The whole equation, stripped of its costume, says something simple: *the future of the cloud is dictated, completely and smoothly, by the cloud's present shape.* Between measurements, there is nothing random about it at all. The wave rolls forward as predictably as a planet. The randomness only crashes in at the moment you look.

That is the great trick and the great scandal of the equation in one breath. The cloud evolves with perfect, clockwork certainty, and yet all it is ever certain about is the *odds*.

The Proof in a Box

Now to the heart of it, and the promise of this book: we will derive those mysterious energy rungs ourselves, the ones that make the colours, using nothing but the idea that matter is a wave. Schrödinger explained quantisation as a kind of music, and that is exactly the path we will walk.

Trap a particle in a small box of length L. An electron held inside an atom is not far off this picture. Now, the particle is a wave, and a wave trapped in a box has to *fit*. Think of a guitar string, pinned down at both ends. It cannot vibrate just any old way. It can only make shapes that are still at both pinned ends: one clean arch, or two, or three, always a whole number of half-waves between the ends. You cannot fit two and a half humps on a guitar string. The ends forbid it. That is why a string of a fixed length sounds only certain notes, and it is the same reason an organ pipe of a fixed length sounds only certain pitches.

The trapped electron wave is under the exact same rule. Only whole numbers of half-waves fit in the box. So the only wavelengths allowed are the ones that divide neatly into twice the box length:

$$\lambda_n = \frac{2L}{n}, \qquad n = 1, 2, 3, \ldots$$

where n is just the number of humps: one, two, three, and so on. No in-between values are permitted. The box simply will not hold them.

Now bring in de Broglie. Every wavelength means a momentum, through $\lambda = h/p$, which rearranges to $p = h/\lambda$. So each allowed wavelength forces an allowed momentum:

$$p_n = \frac{h}{\lambda_n} = \frac{nh}{2L}$$

And finally, momentum means energy. For a moving particle the energy of motion is $E = p^2/2m$, a fact that sits one short step from Newton. Put the allowed momentum in, and out drops the allowed energy:

$$E_n = \frac{p_n^2}{2m} = \frac{n^2 h^2}{8mL^2}$$

Stop and look at what just happened. Because the wave had to fit in the box, the energy can take only the values you get by setting n to a whole number: $1, 4, 9, 16$ times a fixed amount, and never anything between. The ladder of rungs was not put in by hand. It fell out, forced into being, the moment we admitted the particle was a wave squeezed into a small space. The mysterious steps in the atom's energy are nothing more exotic than the harmonics of a plucked string. The atom really is a musical instrument, and it plays only its own notes because only certain waves fit inside it. That is the whole secret of the colours.

Be the Judge

Let us put real numbers in, because this is where it stops being a story and becomes a thing you can check.

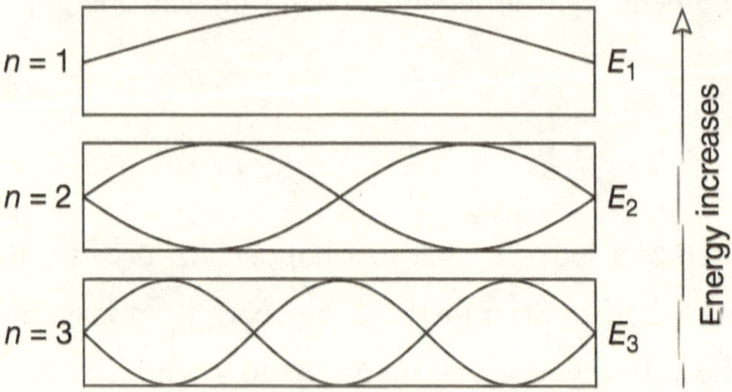

Figure 7: Figure 8.1 – Particle in a box: only whole numbers of half-wavelengths fit, so energy is quantised into rungs.

Take an electron, mass about 9.1×10^{-31} kilograms, and trap it in a box about the size of an atom, a width of roughly 1×10^{-10} metres. Drop those into our formula for the lowest rung, $n = 1$, with Planck's constant $h \approx 6.6 \times 10^{-34}$:

$$E_1 = \frac{h^2}{8mL^2} = \frac{(6.6 \times 10^{-34})^2}{8 \times (9.1 \times 10^{-31}) \times (1 \times 10^{-10})^2}$$

Grind through it on a calculator and you get about

$$E_1 \approx 6 \times 10^{-18} \text{ joules},$$

which is a few tens of what physicists call electron-volts. And that is the punchline: a few tens of electron-volts is exactly the scale of real atomic energies, the very energies that make the colours of the spectral lines. A toy box, a guitar-string argument, and de Broglie's wavelength land you, on your own calculator, right in the middle of real atomic physics.

Now do the same sum for something everyday, and see why we never notice any of this. Put a one-kilogram ball in a one-metre box. The lowest rung comes out at about 5×10^{-68} joules. That is not a typo. The energy steps for the ball are so unimaginably tiny, smaller than the electron's by some fifty zeroes, that no instrument that could ever exist would detect the gaps between them. The ball's allowed energies are so finely spaced they smear into what looks like a perfectly smooth ramp. The ball can sit at any speed at all, as far as anyone could ever tell.

There is the reconciliation of the whole quantum world with the ordinary one, in two sums you can do yourself. The rungs are always there, for everything. They are gigantic and unavoidable for an electron in an atom. They are infinitesimal and invisible for a ball in a room. The quantum world is not a different world. It is this one, seen close enough that the steps finally show.

Why You Exist

Now we can answer the question we opened with. Why does the electron not spiral into the nucleus, taking all of matter down with it?

Because of that lowest rung. Look again at the formula: when $n = 1$, the energy is not zero. It is the smallest the trapped wave is allowed to have, and there is no rung beneath it. The electron cannot give up that last bit of energy and fall further in, because to sit closer to the nucleus it would have to be a wave squeezed into a smaller box, which by our own formula means *more* energy, not less. There is simply nowhere lower to go. The wave will not fit. The bottom rung is a floor, and the electron rests on it forever.

That floor is the reason atoms are stable. It is the reason chemistry exists, the reason matter holds its shape, the reason there is something rather than a flash of collapsing nothing. You are held up, this very second, by the stubborn refusal of a wave to be crushed into too small a space. And every colour you have ever seen, the red of neon, the orange of a street lamp, the precise dark lines astronomers read in starlight to know what distant suns are made of, is an electron hopping between these rungs and paying out the difference as light. The barcode of the elements, the whole palette of the cosmos, is the music of waves trapped in atoms.

The Cat in the Box

There is a price for all this, and it is a deep unease that has never gone away.

If a particle really has no definite position until it is measured, if it truly is a spread-out cloud of maybe, then what counts as a measurement? Where is the line between the ghostly quantum world and the solid one we live in? Schrödinger himself was so troubled by this that he cooked up a thought experiment designed to make the whole idea look ridiculous. It has since become the most famous cat in history.

Imagine, he said, a cat sealed in a box with a fiendish device: a single radioactive atom, and a hammer wired to smash a flask of poison if the atom decays. Whether the atom decays in the next hour is a pure quantum event, governed by Ψ, genuinely undecided until observed. So while the box is shut, the atom is in a blur of decayed-and-not-decayed at once. And if the atom is blurred, then so is everything its fate is tied to. The hammer is fallen and not fallen. The cat is, on paper, both dead and alive, smeared across both possibilities, until somebody opens the box and looks.

Schrödinger meant this as a reductio, a way of saying: see how absurd your interpretation becomes when I scale it up to a cat. The discomfort was the point. But the joke has refused to die, because nobody has ever cleanly explained why it is wrong. The equation says superpositions are real, and they have been demonstrated, beyond doubt, for particles and atoms and ever-larger objects. Where exactly the cloud of maybe stops and the single solid fact begins is a question physicists still argue over today, a full century on. The mathematics is flawless and undisputed. What it *means* is the longest unfinished conversation in science.

The World It Built

Set the philosophy aside, and you find the most practically important equation in this entire book.

Every atom's structure, every way that atoms join into molecules, the whole of chemistry, is the solving of Schrödinger's equation for electrons. The shape of the periodic table, why oxygen grabs at hydrogen, why carbon can build the endless scaffolding of life, all of it is the geometry of these probability clouds deciding how they can share space. Chemistry is just applied Ψ.

And then there is the machine you are likely reading

this on. Every transistor in every microchip works because electrons behave as waves, sliding through layers of engineered material in ways that only quantum mechanics permits. The entire semiconductor industry, every computer and phone and the vast humming farms of servers that run the modern world, is built on equations descended directly from Schrödinger's. So are the lasers in fibre-optic cables and supermarket scanners, the glowing diodes in every screen and indicator light, the medical scanners that map the inside of a living body, the solar cells turning sunlight into power. The strange, ghostly law of the cloud of maybe is, by a wide margin, the most lucrative and load-bearing physics ever discovered. The whole information age sits on top of it. Every signal that has ever crossed a wire began with electrons obeying this equation.

Is It Really True?

It is, perhaps, the most precisely tested theory humanity has. Its predictions have been confirmed in some cases to more decimal places than any other science can manage, and not once, in a hundred years of fierce trying, has a careful experiment caught it out.

The matter-wave idea at its root was confirmed almost at once. Within two years of Schrödinger's papers, experimenters fired electrons at a crystal and watched them ripple and interfere on the far side, fanning into the same banded pattern that water waves and light waves make. Particles really do diffract like waves. Since then the cloud of maybe has been confirmed in ways that still feel impossible: single particles sent one at a time through two slits and interfering with themselves, pairs of particles linked so that measuring one instantly fixes the other no matter how far apart they are. Every strange thing the equation implies has turned out to be how the world actually behaves.

It has one known frontier. Schrödinger's equation, and the whole quantum framework around it, has so far refused to join hands with gravity, the great curved-spacetime picture we will reach soon. The very small and the very heavy are each described to perfection, and the two descriptions will not yet sit in the same room. That unfinished marriage is the largest open problem in physics. But within its own domain, the atom and everything built from it, the equation has never been wrong.

The Name on the Door

We call it Schrödinger's equation, and he did write it. But the door, as ever in this book, has a crowd behind it.

The seed was the Prince, Louis de Broglie, whose wild thesis that matter is wavy was the idea everything grew from. The polished equation was Schrödinger's, found in a now-legendary burst over the Christmas of 1925, when he retreated to a villa in the Swiss Alps and came back down with a theory that historians have called a stretch of creativity without parallel in the history of science. The meaning of the wave, the probability rule that is the soul of the whole thing, was not Schrödinger's at all but Max Born's, who supplied it shortly after and waited nearly thirty years for the Nobel Prize it earned.

And then there is the twist that sets up our next

chapter. Schrödinger's beautiful wave equation was not even the first version of quantum mechanics. Months earlier, a young German named Werner Heisenberg had built a complete quantum theory of his own, by a route so abstract and forbidding that almost nobody could stomach it, full of strange tables of numbers and no waves at all. For a while there were two rival quantum mechanics, looking nothing alike, their supporters quarrelling over which was the truth. Then it was proved that they were not rivals at all. They were the same theory wearing two costumes, giving identical answers by utterly different mathematics. Schrödinger's wave was simply the friendlier face of a thing Heisenberg had already found.

The deepest irony is saved for the man on the door. Schrödinger spent the rest of his life disliking what his own equation had unleashed. He never accepted that the cloud of maybe was the final word, never made peace with a universe that dealt only in probabilities, and dreamed up his poor cat precisely to mock the idea. He built the most successful description of reality ever written, and then refused to believe what it was telling him. Reality, as this book keeps finding, does not need our belief. It only keeps passing the test.

So Look

You cannot watch an electron be a cloud. But you can hold the evidence in your hand tonight.

Look at any coloured light that is not the sun or a flame: a neon sign, the orange sodium glow of an old street lamp, the cool light of an energy-saving bulb. Every one of those colours is electrons dropping between the rungs we derived together, paying out the gap as light of one exact shade. If you can find a cheap prism or even a CD to split that light, you will see it break not into a smooth rainbow but into separate, sharp lines, the fingerprint of the atoms inside, the same fingerprint that lets us read the make-up of stars we will never visit. You are looking straight at the energy ladder, made visible. You are seeing the music of trapped waves.

And the next time you feel how solid the world is, the

table under your hand, the floor under your feet, remember what is really holding it up. Not little balls on little orbits, but waves that refuse to be crushed, resting on the lowest rung they are allowed, exactly as your own calculation said they must.

The equation gave us a universe that does not commit to where anything is. But it still let us calculate the odds with perfect confidence, and that felt like a fair bargain, a strangeness we could live with. Surely, we thought, the blur was only about *position*. Surely, if we were clever and careful enough, we could at least pin down a particle's place and its motion together, and know everything there was to know.

We could not. And the reason why is not a limit of our instruments or our cleverness. It is a wall built into the world itself, a hard and exact boundary on how much can ever be known at once. The next equation does not describe what a particle is doing. It tells us, with brutal precision, what we are permanently forbidden from finding out.

Chapter Nine

Heisenberg's Uncertainty Principle

The Edge of Knowing

$$\Delta x \, \Delta p \geq \frac{\hbar}{2}$$

A little over two hundred years ago, a French mathematician named Pierre-Simon Laplace described a creature that has haunted science ever since. Imagine, he said, a vast intellect that knew, at one instant, the exact position and the exact motion of every single particle in the universe. For such a mind, armed with Newton's laws, the future would hold no mystery whatever. Every tomorrow, every sunrise, every thought you will ever have, would already be written, calculable in advance, as surely as we can predict an eclipse. The universe would be a clock, and if you knew the position of every gear, you could read off all of time.

It is a chilling vision, and for two centuries it looked as though it might be true. Newton's physics seemed to promise exactly this: a world of definite things in definite places moving at definite speeds, knowable in full if only you measured carefully enough. The clockwork might be too vast for any human to compute, but the information was *there*, in principle, waiting.

Then a young German, not yet thirty, proved it was not there. Not hidden, not too small to measure, not waiting for a better instrument. Simply not there. He showed that you cannot know, at the same time, both where a particle is and how it is moving, no matter how clever or careful you are, because the universe itself does not hold both facts at once. Laplace's demon was not merely impossible to build. It was impossible in principle, because the very numbers it would need to read have no definite values to be read. The future is not just hard to predict. It is, at the deepest level, not yet decided. And the equation that says so is one of the shortest and strangest in this book.

The Wall

Here is the whole of it, in symbols you can already half-read:

$$\Delta x \, \Delta p \geq \frac{\hbar}{2}$$

The triangle, Δ, "delta," means "the fuzziness in," the spread, the wobble, the amount you are unsure by. So Δx is the fuzziness in a particle's position, in *where* it is. And Δp is the fuzziness in its momentum, its mass times its velocity, in *how it is moving*. The little $\hbar$ is Planck's constant again, the grain size of the quantum world we met last chapter.

The equation says something brutally simple. Multiply your fuzziness about where a thing is by your fuzziness about how it is moving, and the answer can never, ever drop below a fixed floor. You can know the position superbly well, and then the momentum must go vague. You can pin the momentum down beautifully, and then the position smears out. Push one fuzziness down and the other springs up to compensate, like air in a sealed balloon: squeeze it here, it bulges there. The total can never be squeezed away. There is a hard wall, and behind it the universe keeps its own counsel.

This is not a statement about the limits of our laboratories. It is a statement about the limits of what is true.

It Is Not Clumsy Fingers

There is a popular way of explaining this, and it is wrong, and the wrongness matters so much that we have to clear it out of the way before we can see the real thing.

The popular story goes like this. To find out where an electron is, you have to look at it, and to look at it you have to bounce something off it, a particle of light, say. But an electron is so tiny that the light you bounce off it gives it a kick, and that kick changes its motion. So in the very act of measuring its position, you disturb its momentum. You can never measure one without spoiling the other.

This sounds reasonable, and it is even how Heisenberg himself first tried to explain it. But it is misleading, because it makes the uncertainty sound like a practical nuisance, a problem of clumsy fingers, the kind of thing a cleverer experiment might one day get around. As though the electron really does have a precise position and a precise speed, and we are merely too ham-fisted to read them both without smudging.

That is not what is happening. The deeper truth, the one Bohr pressed on Heisenberg until he accepted it, is far stranger and far more beautiful. The uncertainty is there even when nobody is looking, even when nothing touches the particle at all. It is not that we disturb the values. It is that the values *are not there to be disturbed.* A particle simply does not possess a sharp position and a sharp momentum at the same moment. The information does not exist, anywhere, not even to the universe.

And once you remember what a particle really is, you can see exactly why. Because a particle is a wave. We earned that last chapter. And waves, all waves, even the ordinary ones you can hear, obey this rule already.

The Click and the Note

Let us leave electrons for a moment and listen instead, because the whole secret is hiding in music, and you have heard it a thousand times.

Think of a single, pure musical note. A clean, sustained tone, held steady. To have a definite pitch, that note must go on, and on, evenly, for a good long while. A pure tone is a smooth wave repeating itself many times. The longer it sustains, the more exactly you can name its pitch. A truly perfect, single pitch would have to ring on forever.

Now think of the opposite: a click. A snap of the fingers, a tap on a desk, a sharp tick. It happens at one crisp instant, here and gone. Ask what *pitch* a click is, what musical note, and you cannot answer, because it has none. A click is not a tone at all. If you fed it through the equipment a musician uses to analyse sound, you would find it is a smear of every pitch at once, all jumbled together. The sharper and shorter the click, the wider the smear of pitches inside it.

So here is the trade, and every sound engineer lives with it. A sound that is pinned down precisely in *time,* a click, has no definite *pitch.* A sound with a precise *pitch,* a held note, cannot be pinned down in *time,* it has to last. You can have one or the other, never both perfectly. This is not a flaw in our instruments. It is a property of what waves *are.* Time and pitch are partners locked in exactly this seesaw.

Now bring back the electron, which is a wave. By de Broglie's rule from last chapter, the particle's momentum is set by its wavelength, the spacing of its ripples. So a wave with one single, pure wavelength has one single, definite momentum, the same way a pure note has one definite pitch. But a wave with one pure wavelength is an endless ripple, the same everywhere, stretching across all of space. It is the pure held note. It has a perfect momentum and no location at all. You cannot say where it is, because it is everywhere.

To give the particle a location, you have to build a little bump, a packet, concentrated in one spot and flat everywhere else, so you can point and say "it is roughly *there*." But you cannot make a bump out of a single wavelength. The only way to build a localised bump is to add together many different wavelengths, so that they reinforce in the one spot and cancel everywhere else. That is the click. And a bump built from many wavelengths is a particle built from many momenta. Its momentum is now a smear.

There it is, the whole principle, in plain sound. Position is the *when* of the matter-wave; momentum is its *pitch*. Pin the position into a sharp click and the momentum smears across many pitches. Tune the momentum to one pure pitch and the position spreads out to fill all space. Squeeze the bump narrower for a better position, and you are forced to throw in a wider range of wavelengths, making the momentum vaguer. It is not clumsiness. It is the unbreakable arithmetic of waves, the same one that says a click has no note, written now for the stuff that everything is made of.

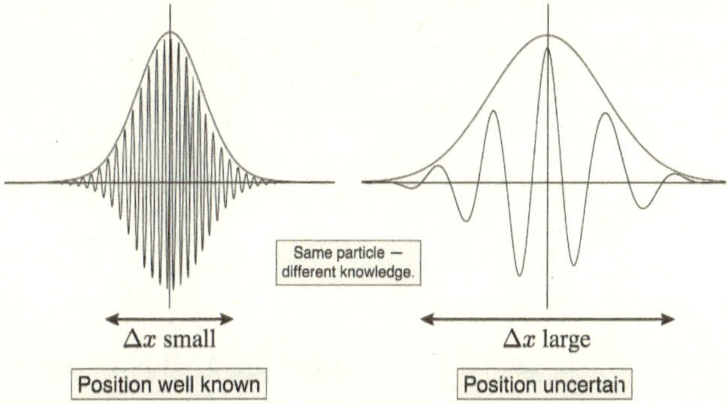

Figure 8: Figure 9.1 — A narrow wave packet has a well-defined position but a spread of momenta; a broad packet has a well-defined momentum but a smeared position.

Be the Judge

Now let us put numbers to it, because the equation does not just gesture at a limit. It tells you the exact size of the wall.

Rearrange it to ask: if I pin a particle's position down to a fuzziness Δx, what is the smallest its momentum fuzziness can be? Simple division gives

$$\Delta p \geq \frac{\hbar}{2\,\Delta x}$$

Take an electron and try to trap it inside a single atom, so its position is fuzzy by only the width of an atom, about $\Delta x \approx 1 \times 10^{-10}$ metres. With $\hbar \approx 1.05 \times 10^{-34}$, the smallest possible fuzziness in its momentum works out to about 5×10^{-25} in the usual units. Divide that by the electron's tiny mass to turn it into a fuzziness in *speed*, and you get a staggering figure:

$$\Delta v \approx 580{,}000 \text{ metres per second.}$$

Read that again. An electron confined to an atom has a built-in fuzziness in its speed of more than five hundred kilometres *per second*. It is not sitting demurely in there. It is irreducibly, permanently aquiver, its motion smeared across an enormous range, and no measurement and no cleverness can ever calm it, because there is no calm value to find. The smaller the box you try to force it into, the more violently uncertain its speed becomes. You can run that division on a calculator yourself.

Now do the same for something from your own world. Take a one-kilogram ball and pin its position to within a single millimetre, a fuzziness most of us would call exact. The equation says its speed must then be fuzzy by at least about 5×10^{-32} metres per second. That is a fuzziness so absurdly, microscopically small that no instrument ever built or ever buildable could detect it. For the ball, the wall is so far away it might as well not exist. You can know its position and its speed together to a precision beyond any practical meaning.

And there, once again, is the bridge between the two worlds. The wall is real and absolute for everything, always. But it stands a colossal distance off for anything large and heavy, and presses right up against the very heart of anything as small and light as an electron. We do not feel uncertainty in daily life for the same reason we do not feel the curve of the Earth in a back garden. It is there. It is just unimaginably gentle at our scale.

Why Atoms Have a Size

This little equation does something its discoverer barely noticed at first: it props up the entire material world, and it tells us, finally and completely, why atoms do not collapse.

Last chapter we found one answer, the lowest rung of the energy ladder. Here is the same truth seen from a deeper angle. Picture the electron's attraction to the nucleus trying to pull it inward, ever closer. But pulling it closer means confining it, shrinking the fuzziness Δx in its position. And the equation is merciless: shrink Δx, and the fuzziness in momentum Δp must balloon. A ballooning momentum means a ballooning energy of motion. So the harder gravity, or rather the electric pull, tries to crush the electron onto the nucleus, the more wildly energetic and unconfinable the electron becomes, and the harder it fights back outward.

An atom is the truce between these two. It settles at exactly the size where the inward pull of attraction is balanced by the outward refusal of a wave to be pinned. That truce distance is the size of the atom, and the size of the atom is the size of everything: the spacing of every brick in every wall, the reach of every chemical bond in every living cell, the very solidity of the chair beneath you. None of it would exist if a particle could be squeezed to a point. The world has bulk, and you can lean on it, because of this one short inequality and its flat refusal to let position and motion both go sharp at once. You are held up, quite literally, by what cannot be known.

The Restless Void and the Wall You Walk Through

The principle has a twin, just as strange, that links energy and time: the more briefly something happens, the less definite its energy can be. And this twin has a consequence that sounds like science fiction and is plain fact.

It means empty space is not empty. For the tiniest slivers of time, energy can be borrowed from nowhere, just enough to flicker particles into being and snuff them out again before the universe can balance its books. What we call the vacuum is in truth a faint, ceaseless seething, a fizz of particles winking in and out of existence everywhere, all the time. This is not a metaphor. Its effects have been measured, in the tiny shifts it makes to atomic energies and in a faint pull it creates between two metal plates held a hair apart. The emptiest emptiness there is turns out to be quietly, restlessly alive.

The fuzziness of position has its own marvel, called tunnelling. Because a particle's position is never perfectly sharp, a particle faced with a barrier it has not got the energy to climb can sometimes simply turn up on the *other side* of it, having never gone over the top. Its cloud of maybe leaks through the wall. This sounds like a cheat, and it is one of the most important facts in nature. It is why the Sun shines. In the Sun's core, the protons that fuse to make its light should not be able to get close enough to fuse at all; they repel each other far too strongly. They fuse anyway, by tunnelling through a barrier they cannot climb. The warmth on your face, which two chapters ago we traced to $E = mc^2$, depends just as much on this: the Sun burns because uncertainty lets its protons walk through a wall. And the same trick, tamed, is at work in the flash memory of your phone and in microscopes delicate enough to feel a single atom.

Is It Really True?

It is among the most secure facts in all of science, and the surest sign of that is how many brilliant people have tried to beat it and failed.

The greatest of them was Einstein. He never made peace with this quantum world of blur and chance. The universe, he felt, must surely have definite values underneath, hidden but real, and he spent years devising ingenious thought experiments to slip past the wall and pin a particle down completely. Each time, Bohr found the flaw, and the wall held. As far as anyone has ever been able to test, and the tests are now exquisite, the fuzziness is not our ignorance of some hidden truth waiting to be uncovered. It is the truth itself. There is no sharper reality underneath to find. But the most famous of those duels, the one night when Einstein came within a hair of winning, deserves to be told in full.

Einstein's Last Trap

It happened in Brussels, in October of 1930, at one of the grand gatherings where the founders of the new physics met to argue it out. Einstein arrived with a machine built in his imagination, the most ingenious challenge to the uncertainty principle he had ever devised, and for one evening it looked as though he had won.

Picture a box, he said, its walls packed with light, and a small shutter in one side worked by a clock sealed inside. We set the clock to flick the shutter open for the briefest instant and snap it shut again, just long enough to let a single photon slip out. Because the clock runs it, we know the *exact moment* the photon left, to whatever precision we please. The time is pinned perfectly. So now, by the uncertainty principle, the photon's *energy* ought to be hopelessly vague.

But it is not, said Einstein, and here the jaws of the trap opened. Weigh the box before the photon leaves, and weigh it again after. The box is now lighter by exactly the mass the photon carried off with it. And by Einstein's own $E = mc^2$ from three chapters ago, that lost mass tells you the photon's energy, as precisely as you can weigh. So you know the energy *and* the moment of escape, both razor sharp at once. The wall is breached. The uncertainty principle, and with it the whole edifice of the new physics, lies broken.

The story goes that Bohr was shaken to his core that evening. Witnesses described him trailing Einstein out of the room, pale and stammering, unable to find the crack in the argument, while Einstein strolled on ahead, faintly amused. For that one night, it was the closest anyone has ever come to toppling quantum mechanics.

And then Bohr sat up through the night, and by breakfast he had it. The flaw was hidden in the single step Einstein had brushed past: *weighing the box*. To weigh a thing you must read its weight off some scale, and that means letting the box settle to a new position, hanging from a spring, say, so a pointer can register how far it sags. The box has to be free to shift, up or down, in the pull of gravity. There is the snare.

The clock that timed the shutter is sealed inside that very box. And by Einstein's *other* masterpiece, his theory of gravity, a clock's rate of ticking depends on its height in a gravitational field. Clocks higher up run a touch faster, clocks lower down a touch slower. We will meet exactly this strangeness in the next chapter.

So the very act of weighing the box, of letting it shift up or down to be read, unavoidably moves the clock inside it through the gravitational field and changes how fast that clock runs, and so blurs the very *moment* the shutter was meant to have caught so perfectly. Grind the numbers through, and the blur it forces into the time, multiplied by the blur in the energy from the weighing, lands exactly on the floor the uncertainty principle demands. Not a fraction less. The wall held. The trap sprang shut on the man who had set it.

And savour the irony, for Bohr surely did. To defend the strange new quantum world that Einstein so distrusted, Bohr reached across and picked up Einstein's *own* theory of gravity, the bending of time itself, and turned it against him. The mightiest argument against the new physics was put down using the mightiest achievement of the man making it. Einstein never again mounted a head-on assault on the uncertainty principle. He fell back on a subtler doubt, that the theory might be entirely correct and still not the whole story, a misgiving that would echo for decades and is not fully laid to rest even now. But on the plain question, whether the wall is real, he had asked, and reality had answered. He could not believe, he said more than once and famously, that God would play dice with the universe. On this he was simply wrong, and he had now been beaten with his own equations. Reality, as this book keeps insisting, gets the final vote, and here it overruled the most powerful mind of the age.

The Name on the Door

Werner Heisenberg arrived at quantum mechanics by a road quite different from Schrödinger's, and he got there first.

In the summer of 1925, a young man of twenty-three and miserable with hay fever, he fled to the bare, pollen-free island of Helgoland in the North Sea to recover. Alone there, he had the idea that broke physics open. He decided to stop imagining what an electron was secretly doing, stop picturing tidy little orbits nobody could ever see, and to build a theory out of *only* the things you could actually measure, like the light atoms give off. What came out was a weird new arithmetic, all grids of numbers, that he could barely interpret himself. His mentor Max Born recognised those grids as a piece of known mathematics called matrices, and together with Pascual Jordan they forged it into the first complete quantum mechanics the world had ever had. It was so abstract, so stripped of any picture you could hold in your head, that many physicists recoiled from it. When Schrödinger's friendlier wave version arrived a year later, the relief was enormous, until, as we have seen, the two were shown to be the very same theory in different clothes. The blurry cloud and the grids of numbers were one thing all along.

In 1927 Heisenberg published the principle that made him a household name. And here the book must puncture the very explanation he led with. In that famous paper, Heisenberg reached for the picture of the disturbing microscope, the story of the kick from the light, the one we spent this chapter dismantling. Bohr, his older colleague, was unhappy with it, and the two argued, sometimes sharply, until Heisenberg came round to the deeper truth that the uncertainty is woven into the wave nature of things and owes nothing to clumsy measurement. So even the man whose name is on the door first told the story wrong, and had to be talked out of it. Even the name itself oversells. The German word he used pointed more at "indeterminacy" than "uncertainty", at values that are genuinely unfixed rather than merely unknown to us, which is closer to the strange truth than the comfortable English word we settled on.

And there is a darker stretch of this door that honesty will not let us walk past. When the Nazis took power and the Second World War came, Heisenberg did not leave Germany, as so many of his colleagues did. He stayed, and he became the scientific leader of the German effort to build a nuclear weapon. What he was truly doing inside that programme has been argued over ever since, and it has never been settled.

Some hold that he tried his utmost and simply failed; some that he quietly dragged his feet to keep the bomb out of Hitler's hands; some that he was never close enough to the answer for his intentions to matter. Recordings secretly made of the captured German scientists after the war show a team that was, in the end, a long way from a working weapon. A tense, mysterious wartime visit he paid to his old mentor Bohr, by then in occupied Denmark, has been turned into plays and books, and no one alive can say for certain what passed between them. I will not pretend to know his heart. I note only that the man who taught the world the limits of what can be known left behind, in his own life, a question that may never be answered.

So Notice

You do not need a laboratory to stand at this wall. You lean on it all day.

The solidity of everything you touch, the floor, the cup, your own hands, is the refusal of matter to be pinned into too small a space. The warmth of the sun on your skin is its protons walking through a barrier they cannot climb. The empty air in front of your eyes is quietly fizzing with particles too brief to catch. And the next time you hear a clean, held note give way to a sharp click, you are hearing, in plain sound, the exact law that governs the heart of every atom: that to be sharp *here* is to be blurred *there,* and that nothing in the universe is allowed to be perfectly both.

It is tempting to feel this as a loss, a door closed on us, a limit to our reach. It is the opposite. This blur is not a defect in the world. It is what gives atoms their size and the world its substance, what lets the stars burn and empty space hum with quiet energy. We are not robbed by the edge of knowing. We are built on it.

We have now mapped the world of the very small as far as this book will go: the cloud of maybe, and the wall around what can be known. The quantum revolution tamed the atom and everything made from it. But all through these two chapters, one ancient giant has stood untouched at the edge of the room, the very first force any of us ever felt, the one that pinned us to the ground before we could walk. Gravity. Newton had described its pull with breathtaking accuracy, but had confessed he had no idea what it *was*. The same restless mind that gave us $E = mc^2$ would spend ten hard years on that question, and the answer, when it came, was the most astonishing in all of physics: that gravity is not a force at all, but a bending of space and time themselves. And lurking underneath it waits the greatest unsolved problem we have, the fact that the quantum world we just explored and the gravity we are about to meet still, to this day, refuse to speak to one another.

Chapter Ten

Einstein's Field Equations

The Shape of Space and Time

$$G_{\mu\nu} = \frac{8\pi G}{c^4} T_{\mu\nu}$$

You think you know what gravity is. You can feel it this very moment, holding you down in your seat, the steady pull that has anchored you to the ground since before you could walk. It is the most familiar force in the world, the first physics any of us ever learns, the reason things fall.

The most successful theory of gravity ever written says that feeling is a lie.

There is no force pulling you down into your chair.

Nothing is tugging on you at all. According to the equation at the head of this chapter, you are not being pulled, you are being *pushed*, upward, by the chair, and the only reason you do not float off is that the chair stands in your way. Left alone, with the chair removed, you would not be yanked downward by some invisible hand. You would simply coast, freely and weightlessly, in a perfectly straight line, through a space that has been bent. Gravity, the theory says, is not a force. It is the shape of space and time themselves, and everything that falls is just travelling as straight as it can through a bent world.

This is the strangest idea in this book, and quite possibly the most beautiful. It cost the man who found it ten years of the hardest thinking of his life, it overturned an understanding that had stood unchallenged since Newton, and it has since been confirmed by everything from the satnav in your phone to the collision of black holes a billion light-years away. I left the last chapter on a clue: that gravity could bend the flow of time, and that Bohr had used the fact to win an argument. It is time to take that clue seriously, all the way down.

The Hole in Newton's Masterpiece

Cast your mind back to Chapter Three. Newton gave us a law of gravity so good it ran the heavens for two centuries, predicting eclipses, guiding spacecraft, even turning up a new planet on the tip of a pen. But Newton himself, for all his triumph, confessed there was something he could not explain. His law said the Sun reaches across a hundred and fifty million empty kilometres and pulls on the Earth. But *how*? Through what? Across sheer nothing? The idea of one body yanking another across empty space, instantly, with no rope and no contact and nothing in between, struck even Newton as absurd. *I frame no hypotheses,* he wrote, and left the question hanging for anyone who came after.

There was also a tiny, nagging flaw. The planet Mercury, swinging closest to the Sun, traces an orbit that slowly rotates, its nearest point creeping around the Sun over the centuries. Most of that creep, astronomers could explain as the tug of the other planets. But a sliver remained, a stubborn 43 arcseconds per century, a tiny angle, that Newton's law simply could not account for, no matter how carefully anyone calculated. It was the same astronomer who had found Neptune, Le Verrier, who pinned down the discrepancy. For half a century it sat there, a hairline crack in the most successful theory in the history of science.

A perfect description, with no explanation, and one small thing it got wrong. That was where gravity stood when a patent clerk turned his attention to it.

The Happiest Thought

The whole revolution grew from a single image that came to Einstein in 1907, one he later called the happiest thought of his life.

He imagined a man falling from a roof. And he realised something obvious that nobody had taken seriously: the falling man feels no gravity. As he falls, if he lets go of his keys, they do not drop away from him, they hang there beside him, falling at exactly the same rate. For the falling man, gravity has simply vanished. We have all seen this now, in footage of astronauts. They are not beyond gravity's reach up there; the Earth pulls them strongly. They float because they are *falling*, endlessly, around the Earth, and a body in free fall feels no weight at all. Falling cancels gravity, perfectly, completely.

Now turn the thought around, which is where it becomes deadly. Imagine you wake in a sealed room with no windows. You feel your normal weight, feet pressed to the floor. Where are you? You might be at rest on the surface of the Earth, held by gravity. Or you might be deep in empty space, far from any planet, in a room being hauled "upward" by a rocket, accelerating at just the right rate, so that the floor presses up into your feet exactly as the ground does at home. And here is the point: *there is no experiment you could do, sealed in that room, to tell the two apart.* Drop a ball, and in both cases it falls to the floor in just the same way. Standing in gravity and accelerating through empty space are not merely similar. They are, as far as physics is concerned, the very same thing.

This is the equivalence principle, and it is the seed of

everything that follows. Gravity is acceleration. Acceleration is gravity. They are two names for one experience. And once you believe that, you can work out the consequences with nothing more than a torch and a careful imagination, and the consequences are astonishing.

First Consequence: Light Falls

Take that sealed room, out in space, and accelerate it "upward" with a rocket. Now shine a torch horizontally across it, from one wall to the other.

While the light is crossing, the room is accelerating upward beneath it. So by the time the beam reaches the far wall, the wall has rushed up a little to meet it, and the light lands slightly *lower* on the far side than the height it set out from. To you, inside the room, the beam of light visibly bowed downward on its way across, curving toward the floor.

Now apply the equivalence principle, the rule we just earned. If this happens in an accelerating room, it must happen, identically, in gravity, because the two cannot be told apart. Therefore gravity bends light. A beam of light passing through a gravitational field must curve, falling like everything else.

This was a stunning prediction, because light, having no mass, was the one thing Newton's gravity should barely touch. Einstein said it would bend, and he said by exactly how much: starlight grazing the edge of the Sun should be deflected by 1.75 arcseconds, twice the feeble amount old physics could muster. The trouble was that you could never see stars next to the Sun, drowned as they are in its glare. Unless the Sun went dark.

In 1919, the British astronomer Arthur Eddington sailed to an island off Africa to photograph the stars around the Sun during a total eclipse, when the Moon blacked out the glare for a few minutes. The stars near the Sun's edge had shifted from their usual positions, pushed outward, by just the amount Einstein had predicted, and twice what Newton allowed. When the result was announced, it made headlines around the entire world. *Lights All Askew in the Heavens.* Newton overthrown. Overnight, a German physicist most people had never heard of became the most famous scientist alive, the public face of genius itself. (Historians have since argued about whether Eddington's 1919 photographs were really sharp enough to settle it; but later measurements, far more precise, have confirmed Einstein's figure beyond any doubt.)

Second Consequence: Time Slows

The deepest consequence is stranger still, and it is the one you can check for yourself, because you carry the proof in your pocket.

Back to the accelerating room. This time, fix a light source on the floor, sending pulses up to a detector on the ceiling, like a tiny clock ticking out flashes. While each flash travels up, the ceiling is accelerating away from it, racing off ahead. So the flashes arrive at the ceiling stretched out, spread further apart in time than they were sent, their rhythm slowed, their colour reddened. The ceiling sees the floor's clock ticking *slow*.

Apply the equivalence principle once more. What is

true in the accelerating room must be true in gravity. So in a gravitational field, a clock lower down, deeper in the pull, runs slower than a clock higher up. Not because the clocks are faulty. Because *time itself* runs at different rates at different heights. Down where gravity is stronger, time genuinely flows more slowly. Your head, right now, is ageing a shade faster than your feet. It sounds like madness. It is measured fact, confirmed with atomic clocks carried up a single flight of stairs.

And here is where it stops being a curiosity and starts running your daily life. The satellites of the GPS network orbit some twenty thousand kilometres up, where Earth's gravity is much weaker, so their clocks tick *faster* than clocks on the ground, by about 45 millionths of a second every day. (Their orbital speed drags this back by about 7 millionths, through the time-stretching of the last relativity; the two effects do not cancel.) The net result is that each satellite's clock gains around 38 millionths of a second per day over yours.

Thirty-eight millionths of a second sounds like nothing. But GPS works by timing radio signals, and in 38 millionths of a second a radio signal travels about 11 kilometres. So if the system ignored Einstein, your position would drift by roughly ten kilometres every single day. Your phone would place you in the wrong suburb by lunchtime and the wrong city by the weekend. The whole network would be useless within minutes of switch-on. The engineers knew this. Every GPS satellite is launched with its clock deliberately set to run at the "wrong" rate, slowed before lift-off by exactly the amount Einstein's equations demand, so that once it is up in weaker gravity, it ticks in step with the ground. Every time your phone finds you on a map, it is quietly trusting that gravity slows time, and proving Einstein right, again, for the millionth time that day. You can hold that fact in your hand.

What Gravity Really Is

Pull these threads together and the new picture of gravity stands revealed, and it is nothing like a force.

Mass and energy, the theory says, bend the very fabric

of space and time around them, a four-dimensional fabric we call spacetime. And things moving through that bent fabric simply follow the straightest path available to them, the way a ball rolled across a hilly field curves not because anything pulls it sideways but because the ground beneath it is sloped. The Earth does not reach out and grab the Moon. The Earth's mass curves the spacetime around it, digging a kind of valley, and the Moon merely rolls along the straightest track it can find in that valley, going round and round. The apple does not get pulled down. The apple coasts in a straight line through a space that the Earth has bent toward itself. There is no force. There is only shape.

The picture people reach for is a bowling ball sitting on a stretched rubber sheet, dimpling it, so that marbles rolled nearby spiral inward. It is a useful image, and you have probably seen it. But be warned, because it is also misleading, and almost every telling leaves the deepest part out. The rubber sheet shows space being bent. But for everyday falling, for the apple, for you, it is not mainly *space* that is bent at all. It is *time*. Remember what we just derived: clocks

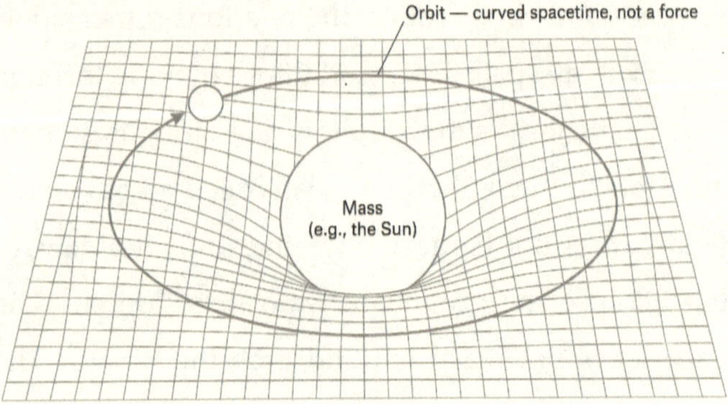

Figure 9: Figure 10.1 – Spacetime curvature: mass bends the fabric of spacetime, and everything else follows the straightest path through the bend.

run slower lower down. An object thrown into the air follows the path through spacetime that keeps it ageing as much as possible, and because time runs slower nearer the Earth, that path bends back toward the ground. You fall not because space is curved like a trampoline, but because, down here, time itself runs slightly slow, and your natural path through spacetime tips you toward the slow side. Gravity, for slow-moving things like us, is very nearly the same thing as the slowing of time we just put in your pocket. The physicist John Wheeler summed the whole theory in a single line: spacetime tells matter how to move, and matter tells spacetime how to curve.

Reading the Equation

That single line is exactly what the equation says, in the dense shorthand of mathematics:

$$G_{\mu\nu} = \frac{8\pi G}{c^4} T_{\mu\nu}$$

Take it in two halves. On the right, the term $T_{\mu\nu}$ is a complete accounting of all the matter and energy at a place: how much there is, how it is moving, how it presses and flows. It is the *stuff.* On the left, $G_{\mu\nu}$ describes the curvature of spacetime there: how steeply the fabric is bent, in every direction. It is the *shape.* The middle, $8\pi G/c^4$, is just a bundle of constants setting the exchange rate, with Newton's old gravitational constant G sitting inside it, and the speed of light c telling you, by its sheer size, how stiff spacetime is, how much matter it takes to bend it even a little.

So the whole equation reads, simply: *the shape of space-*

time here equals the matter and energy here. Stuff on the right, shape on the left, welded together as one truth. Put mass and energy into the universe, and spacetime must curve to match. That is gravity, complete.

But notice those little subscripts, the μ and ν. They are a warning. They mean this is not really one equation at all. It is a tightly woven bundle of ten separate equations, each describing the curvature along different combinations of the dimensions of space and time, and they are all tangled together, each feeding into the others. They are, by common agreement, among the most ferociously difficult equations in all of physics. The simple, godlike sentence on the page hides a thicket so dense that Einstein, one of the deepest minds who ever lived, needed help to find his way through it, and spent a decade lost inside.

The Ten-Year Climb

The decade nearly broke him. Einstein had the central idea, the equivalence principle, by 1907. It took him until the end of 1915 to turn it into the equation above, and the reason it took so long is that the mathematics of curved spaces is monstrous, and Einstein, for all his physical genius, did not know it. He had to learn it, and he leaned heavily on an old university friend, the mathematician Marcel Grossmann, begging him for help: *Grossmann, you must help me or I shall go crazy.* Grossmann pointed him to a strange and beautiful geometry of curved, many-dimensional spaces that had been worked out, purely for its own sake, decades earlier by Bernhard Riemann, a geometry nobody had imagined could describe anything real. It had been sitting on a shelf for half a century, waiting. We have seen this before, with Euler: mathematics invented for no reason at all, turning out to be the secret language of the universe.

For years Einstein wrestled with it, found promising equations, convinced himself they were wrong, threw them out, circled back. By late 1915 he was racing, partly against the great mathematician David Hilbert, who had grasped what Einstein was attempting and was closing in on the same equations from the mathematical side. In November 1915, in a fever of work, Einstein found the final form.

And he tested it at once, on the oldest open wound in the theory of gravity: Mercury. He fed the planet's orbit into his new equations, with nothing adjusted, nothing fudged, and out came the missing piece, the stubborn 43 arcseconds per century that had defied Newton for fifty years, predicted exactly, falling out of the pure geometry of bent spacetime. Einstein later said that when he saw it, he was so overcome that he felt something snap inside him, that he had heart palpitations and could not work for days. He had not invented a number to fit the data. The number had simply appeared, demanded by the shape of space and time. The crack in Newton's masterpiece was sealed. Reality had voted, and it had voted for him.

The Things It Predicted

Here is the mark of a truly great equation: it knows more than the person who wrote it. Einstein's equations went on to predict, all on their own, a series of things so outlandish that Einstein himself disbelieved several of them. Every one has since been found to be real.

The equations said that if you crushed enough mass into a small enough space, the curvature would run away to infinity, digging a pit in spacetime so steep that not even light could climb out. A region of the universe sealed off forever, from which nothing returns. A black hole. Einstein thought such things could not really exist in nature. They do. Our galaxy has a giant one at its centre, and in 2019 astronomers, linking telescopes across the whole Earth, took an actual photograph of the glowing rim around one in a distant galaxy, a ring of light circling a pit of pure darkness. A few years later they imaged the monster at the heart of our own. The pictures match Einstein's geometry exactly.

The equations also said, when first applied to the whole universe, that the cosmos could not sit still. It had to be either expanding or contracting; a static universe was forbidden. Einstein found this so distasteful that he added an extra term to his own equations by hand, a fudge called the cosmological constant, just to hold the universe still. Then, in 1929, Edwin Hubble looked through his telescope and found that the universe *is* expanding, the galaxies all rushing apart, exactly as the untouched equations had insisted. Einstein had had the discovery of the expanding universe in his grasp and flinched. According to the physicist George Gamow, he later called that fudge the greatest blunder of his life. The deepest irony is that the term has since come roaring back, because the expansion of the universe is speeding up, driven by something we call dark energy, and Einstein's "blunder" may describe it after all.

And the equations predicted that violent events, like two black holes colliding, would send ripples through the fabric of spacetime itself, waves of pure bending, spreading outward at the speed of light and stretching everything they pass through. Einstein predicted these gravitational waves in 1916, and was sure they were far too faint to ever detect. For a century he looked right. Then, on the 14th of September 2015, two enormous detectors in America felt a wave wash through the Earth, a tremor in spacetime from two black holes that had spiralled together and merged more than a billion years ago, each around thirty times the mass of the Sun. In the final instant of that collision, the power poured out as gravitational waves briefly outshone every star in the entire visible universe combined. By the time it reached us, spread thin across a billion light-years, it stretched the detectors by less than one thousandth of the width of a single proton, and we caught it. A century after one man predicted it with a pen, we heard two black holes collide.

Is It Really True, and the Crack at the Bottom

It is among the most precisely tested and triumphant theories in all of science. It has passed every experimental test ever devised, from the bending of starlight to the ticking of the clock in your phone to the ringing of spacetime after a black-hole collision. Within its domain, it has never once been wrong.

And yet, in the honest spirit of this book, it cannot be the final word, and the reason is the deepest unsolved problem in physics. I have now told both of the twentieth century's great stories. The quantum world of the last two chapters, where the very small comes in lumps and blurs into probability. And the relativity of this one, where gravity is the smooth, seamless bending of spacetime. Each is fantastically accurate. Each has passed every test. And they flatly contradict each other. General relativity needs spacetime to be a smooth, continuous fabric. Quantum mechanics says everything, examined closely enough, is grainy, uncertain, jittering. In almost every situation this clash never surfaces, because gravity is too weak to matter for tiny particles and quantum fuzziness is too small to matter for planets. But in the two places where something is both extremely massive and extremely tiny, at the very centre of a black hole, and at the first instant of the universe, both theories apply at once, and they scream different answers, and the mathematics dissolves into nonsense. Somewhere there must be a deeper theory, a theory of quantum gravity, that contains them both and heals the contradiction. Finding it is the great unfinished business of physics. The two most successful ideas we have ever had will not sit in the same room, and no one yet knows the words that will make them shake hands.

The Name on the Door

This door, unusually, belongs almost wholly to one man. Most of the equations in this book were team efforts smeared across decades and credited to whoever happened to be standing nearest. General relativity is the rare case of a single mind building a whole new universe nearly alone, over ten years, from one stubborn thought about a falling man. The name belongs on the door.

But not the door entirely. There was Marcel Grossmann, who handed Einstein the geometry without which he would have stayed stuck, and Bernhard Riemann, long dead, who had built that geometry out of pure imagination half a century before anyone knew it was the shape of reality. And there was David Hilbert, perhaps the greatest mathematician of the age, who was chasing the same equations through November 1915 and arrived at the field equations within days of Einstein, by a more purely mathematical route. For a while there was an awkward question of who had got there first. It was settled with grace: Hilbert himself never claimed the theory, insisting it was Einstein's in every way that mattered, the physics, the vision, the decade of struggle. The equation could be reached from the mathematics in a week by a genius who knew where to look. The universe behind it took Einstein ten years to see.

So Look Up

You do not need an eclipse or a black hole to stand inside this equation. You are inside it constantly.

Your phone finds you on a map only by trusting that gravity slows time. Your head is ageing imperceptibly faster than your feet, as you read this. The sunlight on your skin set out from a star whose own light bends the paths of the stars behind it. The ground holds you up not by pulling, but by blocking your free fall through a universe that the Earth has gently bent. The most familiar sensation you have, your own weight, is not a force at all. It is the feeling of the floor interrupting your natural, straight-line coast through curved spacetime. Gravity is not something the universe *does*. It is the shape the universe *has*.

We have now climbed about as high and dived as deep as physics can take us, out to the bending of space and time, and down into the lumpy blur of the atom. We have met the laws that govern motion, light, matter, gravity, and the quantum. And every single one of them, from Newton to Einstein to Schrödinger, shares a quiet and very strange feature that we have not yet remarked upon. They all work exactly the same whether time runs forwards or backwards. Not one of these great laws cares about the direction of time. Run the equations in reverse and they are just as happy; a film of two colliding billiard balls, or two merging black holes, looks perfectly lawful played backwards.

And yet you know, as surely as you know anything, that time runs one way only. A dropped cup shatters; it never leaps back together. We grow old; we never grow young. The past is fixed and gone, the future open and unreached. Where, in a set of laws that do not care about the direction of time, does the relentless one-way arrow of time come from? The answer is the next equation, and it is the only law in this book with time's arrow built into its very heart. It is about heat, and disorder, and the slow running-down of everything, and it may be the most quietly profound law of all.

Chapter Eleven

Second Law of Thermodynamics

The Arrow of Time

$$\Delta S \geq 0$$

Watch a film of a wine glass falling from a table and shattering on the floor. Now watch it run backwards. The shards leap up off the tiles, gather themselves into a whole glass, and settle gently back on the table's edge. You laugh, because it is absurd. You have never seen it happen, and you know in your bones that you never will.

But here is the thing that should stop you cold. Every law of physics in this book so far would happily allow it. Newton's laws, Maxwell's, Schrödinger's,

Einstein's, every one of them works exactly the same whether time runs forwards or backwards. Play a film of the planets orbiting backwards, and they obey gravity perfectly. Run the collision of two billiard balls in reverse, and not a single law is broken. At the level of the deep equations, the future and the past are mirror images, interchangeable, with no arrow pointing either way.

And yet time has an arrow, the most obvious fact of your entire existence. Eggs scramble but never unscramble. Heat spreads from the hot coffee into the cool room, never the other way. You remember yesterday and not tomorrow. You were born, you grow old, and the reverse will never happen. Of all the things you are certain of, none is more certain than that time runs one way only. So where does that arrow come from, if not one of the great laws contains it?

It comes from a single equation, the only one in this book with the direction of time built into its heart. It is not about forces, or particles, or fields. It is about something far humbler and far stranger: counting. And it reveals that the relentless one-way flow of time, the ageing of your face, the cooling of your tea, the whole irreversible river of your life, is, at bottom, nothing but a matter of overwhelming odds.

The One Law That Knows Which Way Time Goes

The equation is the second law of thermodynamics, and in its sharpest form it is almost insultingly simple:

$$\Delta S \geq 0$$

It says that a certain quantity, S, called entropy, can only ever stay the same or increase, in any system left to itself. It can never, of its own accord, go down. The Δ, "delta," means "the change in," so the whole law reads: the change in entropy is greater than or equal to zero. Entropy never decreases.

That is the whole arrow of time, in three symbols. The past is the direction of lower entropy. The future is the direction of higher entropy. Time flows the way entropy grows. Every other law in physics shrugs at the difference between forwards and backwards. This one does not. This one points.

But to feel why it is true, and why it is irreversible when nothing else is, we have to understand what this strange quantity entropy actually is. And the usual one-word answer, "disorder," is a half-truth that hides the real and far more beautiful idea underneath.

What Entropy Really Is

People will tell you entropy is messiness, and that the second law says things tend to get messy. It is a serviceable picture, but it is not quite right, and the precise version, which we owe to Ludwig Boltzmann, is one of the deepest ideas humans have ever had.

Entropy is about *counting*. Specifically, it counts the number of different microscopic arrangements that all look the same from the outside.

Think of your bedroom. There is exactly one arrangement of everything that counts as "perfectly tidy", every book on its shelf, every sock in its drawer.

But there are countless billions of arrangements that count as "messy", clothes here, books there, the exact scatter barely mattering. "Tidy" is a state you can make in very few ways. "Messy" is a state you can make in an astronomical number of ways. That is the entire secret. A low-entropy state is one there are *few* ways to make. A high-entropy state is one there are *vast numbers* of ways to make.

And now the second law stops being a mysterious commandment and becomes something close to obvious. Things drift from order to disorder not because some force pushes them, but because there are so overwhelmingly many more disordered arrangements than ordered ones that a system left to shuffle itself blindly is almost certain to end up in one of the countless messy states rather than one of the rare tidy ones. It is not a law of physics in the way gravity is. It is a law of *counting*. Disorder wins because disorder is more common. That is all.

Let us prove it to ourselves, because, as ever, this book would rather show you than tell you.

The Proof in a Handful of Coins

Take four coins and toss them. Each lands heads or tails, so there are sixteen possible outcomes in all, and every single one is equally likely; the coins have no preference.

Now group those sixteen outcomes by what they look like from a distance, by how many heads came up. Count carefully:

- All four heads: there is exactly **1** way to do it.
- Three heads, one tail: **4** ways (the tail can be any of the four coins).
- Two heads, two tails: **6** ways.
- One head, three tails: **4** ways.
- All four tails: **1** way.

Look at what the counting reveals. The neat, lopsided result, all four heads, can happen in only one way. The balanced, mixed result, two and two, happens in six. So an even split is six times more likely than all-heads, and nothing caused that except the raw number of ways each can occur. The mixed state is not preferred by any force. It is simply more numerous.

Now scale up, because scaling up is where it becomes the iron law of the universe. Toss not four coins but a hundred. The number of possible outcomes is now two multiplied by itself a hundred times, a number with thirty digits. And of all those, the perfectly ordered "all hundred heads" is still just **one**. The arrangements clustered around fifty-fifty number in the hundreds of thousands of trillions of trillions. If you tossed a hundred coins once every second since the beginning of the universe, you would never, not once, have seen them all come up heads. Not because it is forbidden. Because it is drowned, utterly, by the sheer count of messier possibilities.

This is Boltzmann's insight, and he carved it into an equation of breathtaking economy:

$$S = k \log W$$

Here W is exactly the thing we have been counting: the number of microscopic ways a state can be arranged. The log tames that monstrous number into something manageable, and k is a tiny constant, now called Boltzmann's constant, that sets the units. That is all entropy is: a measure, through W, of how many ways. And the second law, $\Delta S \geq 0$, is now no mystery at all. A system wanders blindly among its possible arrangements, and since there are so many more high-W arrangements than low-W ones, it drifts, with overwhelming and unstoppable probability, toward the states there are more ways to be in. Toward higher entropy. Always. The arrow of time is the universe, every instant, falling into its more probable states, simply because they are more probable.

Be the Judge

Put numbers on it and feel the size of the thing.

The chance of those hundred coins all landing heads is one half, multiplied by itself a hundred times: about 8×10^{-31}. That is a decimal point followed by thirty zeros and then an 8. You can punch one half, to the power of any number you like, into a calculator and watch how savagely fast the chance of order collapses as the number grows.

Now stop thinking about coins and think about the air in the room you are sitting in. It contains something like a thousand trillion trillion molecules, all jostling at random. There is no law forbidding them, at some instant, from all happening to crowd into one half of the room, leaving you gasping in vacuum on the other side. It is perfectly allowed. Its probability is one half raised to the power of that thousand-trillion-trillion, a number so close to zero that you could wait through the lifetimes of a billion billion universes and never glimpse it. That is *why* the air stays evenly spread. Not because a force holds it there. Because the spread-out arrangements outnumber the crowded ones beyond all conceivable counting.

Here is the one you can do tonight. Take an ordinary deck of fifty-two cards and shuffle it well. The number of possible orderings of a deck is fifty-two multiplied by fifty-one by fifty by all the way down to one, and it comes out to roughly 8×10^{67}, an eight followed by sixty-seven zeros. That number is so vast that it dwarfs the number of atoms in our galaxy. Which means this: the particular order your shuffled deck is now in has, almost certainly, never existed before in the entire history of card-playing, anywhere on Earth, in all the billions of shuffles ever made. You are holding an arrangement that is, with near-perfect certainty, brand new to the universe. Every careless shuffle creates something unique. *That* is how few the ordered states are, and how unthinkably many the disordered ones. That is the second law, in your hands.

The Law That Came From Steam

For all its cosmic reach, this law was not born from contemplating the universe. It was born in the soot and clamour of the Industrial Revolution, from people trying to build a better steam engine.

In 1824, a young French engineer named Sadi Carnot asked a practical question: how much useful work can you possibly squeeze out of a given amount of heat? And he found a hard and surprising limit. You can never turn all of the heat into work. No matter how perfect your engine, some of the heat always escapes, spilled away, unusable. Heat will flow on its own from hot things to cold things, and you can harness a little of that flow to do work, but you can never run it perfectly, and you can never make it flow backwards, from cold to hot, without paying for it with even more effort spent elsewhere.

This is why there is no perpetual motion machine, and never will be. It is why your phone gets warm in your hand, why every power station dumps waste heat into a river or the sky, why no engine ever built or ever buildable is perfectly efficient. Some energy always degrades into useless, spread-out warmth. Rudolf Clausius, building on Carnot, saw that this "spreading and degrading" was a single universal tendency, gave the quantity that measures it a name, *entropy*, from the Greek for transformation, and stated the law that would outlast every engine: the entropy of the universe tends always toward a maximum. The practical rule of the workshop turned out to be a law of all creation.

The Long Running-Down

Follow the law to the end of time and it tells a sober story.

If entropy can only increase, then the universe is winding down, like a clock whose spring is slowly releasing. The stars are burning through their fuel. Heat is flowing, relentlessly, from the hot places to the cold, evening itself out. And the equations point toward a far-distant future when this process has run all the way to completion: when every star has died, every difference in temperature has been smoothed away, and the entire cosmos has settled into a uniform, lukewarm, featureless sameness. A universe at perfect equilibrium, with no hot and no cold, no flow, no usable energy left anywhere. And with no differences left to exploit, nothing can ever happen again. No engine can turn, no star can shine, no thought can form. Physicists call it the heat death of the universe, and it is the price of the arrow of time. The same law that lets anything happen at all guarantees that, in the end, everything stops.

And Yet, Here We Are

If everything slides toward disorder, you might well ask how *you* are possible. A living body is one of the most exquisitely ordered things in the universe, billions of cells in precise arrangement, sustaining themselves, building, growing, thinking. How does life, this island of stunning order, arise in a universe sworn to decay?

The answer is one of the most beautiful ideas in all of science, and it does not break the law for a moment. The second law forbids entropy from falling in a *closed, isolated* system, one that exchanges nothing with the outside. But you are not isolated. You are an open system, with a river running through you. You take in low-entropy order, in the form of food and oxygen, and you pour out high-entropy waste and heat. You build and maintain your own astonishing internal order by *dumping even more disorder into your surroundings* than the order you create. The books always balance, and they always balance in entropy's favour. Life does not defy the second law. It is a whirlpool in the great downstream current of it, a local eddy of order that exists only because it speeds the overall slide toward disorder.

And what powers the whole arrangement? The Sun.

The Earth drinks a steady stream of low-entropy sunlight, that concentrated gift of energy we traced three chapters ago to $E = mc^2$, and radiates back out into the cold dark of space a far larger flood of high-entropy heat. Life on Earth lives entirely on the difference, on that one-way flow from the brilliant hot point of the Sun to the blackness around it. Every ordered thing you have ever seen, every leaf, every creature, every city, is paid for by the Sun's slow burning and the universe's slow running-down. Schrödinger himself, the man of Chapter Eight, wrote a small and famous book asking *What Is Life?*, and his answer was exactly this: a living thing survives by feeding on order, by drinking in low entropy faster than it makes its own. We are not exceptions to the decay of the universe. We are some of its most beautiful consequences.

Why the Arrow Points That Way

There is one last puzzle, and it is a deep one. If the underlying laws do not care about the direction of time, then entropy should be just as likely to be higher in the past as in the future. Why, then, does it only ever climb toward tomorrow?

The answer, as far as we can tell, lies all the way back at the beginning, in the chapter before this one. The universe must have started, at the Big Bang, in a state of extraordinarily *low* entropy, almost unimaginably ordered. And ever since, for nearly fourteen billion years, it has been doing the only thing the counting allows: sliding downhill, toward more probable, more disordered states. The arrow of time is not really a property of the laws at all. It is an inheritance. It points away from the strange, pristine, low-entropy moment of creation, and it will keep pointing that way until the running-down is complete. Why the universe began so ordered, nobody yet knows. It may be the deepest unanswered question we have. But the whole one-way river of time, your whole sense of past and future, flows from that single, mysterious, beautifully ordered first instant.

Is It Really True?

No law in all of science is held in higher regard, or has survived more attempts to break it. The great astronomer Arthur Eddington, who gave the arrow of time its very name, put it most fiercely: if someone shows you a theory that disagrees with Maxwell's equations, well, so much the worse for Maxwell; if it disagrees with experiment, experiments are sometimes wrong. But if your theory is found to go against the second law of thermodynamics, he said, there is no hope for it; it can only collapse in deepest humiliation. Every claimed perpetual motion machine, in two centuries of hopeful tinkering, has failed. Every one. The law has never once been caught out. It is, by wide agreement, the surest thing we know about the physical world.

The Name on the Door

Three names belong here, and the third carries a tragedy.

There is Sadi Carnot, the young engineer who started

it all in 1824 with his study of heat engines, and who died of cholera just eight years later, at thirty-six, his work nearly forgotten until others revived it. There is Rudolf Clausius, who turned Carnot's workshop insight into a universal law and gave entropy its name. And there is Ludwig Boltzmann, who saw deepest of all, who understood that entropy was about counting arrangements, that the second law was really a law of probability, and who tied the visible world of heat to the invisible swarming of atoms with the equation $S = k \log W$.

For that vision Boltzmann paid dearly. In his lifetime, many of the most powerful physicists of the day refused to believe in atoms at all, dismissing them as a convenient fiction, since no one had ever seen one. Boltzmann's entire framework depended on atoms being real, and he found himself fighting, year after year, bitter and personal battles to defend it, increasingly feeling like the last man standing. He also struggled all his life with a darkness that had nothing to do with physics, with severe depression and failing health, his moods swinging violently between elation and despair. In 1906, while on holiday with his family near Trieste, he took his own life. Within just a few years, the work of Einstein and the experiments of Jean Perrin proved beyond any doubt that atoms are real, and Boltzmann's life's work was triumphantly vindicated. He did not live to see it. And there is one final, quiet irony, perfectly in keeping with this book: the famous equation $S = k \log W$ carved on his tombstone in Vienna was never written in quite that form by Boltzmann himself. Even his own monument bears an equation finished by other hands.

So Notice

You do not need an experiment to meet this law. You are swimming in it.

The arrow of time is the most intimate fact of your life. It is why your coffee cools and never reheats itself, why you remember your childhood and not next year, why the broken glass stays broken. It is the reason you are young only once. Every breath you take, every meal you eat, is you holding back the tide for a while, building a little order at the cost of a little more disorder spilled into the world, running on the Sun's slow fire. You are a brief, beautiful eddy in a river that flows one way only, from the ordered birth of the universe toward its quiet, level end.

And now, to close this book, comes the most surprising turn of all. We have spent this chapter saying that entropy measures the number of arrangements you cannot distinguish, the amount you do not know about the precise state of things. Sit with that phrasing. Entropy is a measure of *missing knowledge.* Of *information* that is absent. And it turns out, astonishingly, that this is not a loose metaphor. The disorder of a box of gas and the data on your phone are governed by the very same mathematics. The last two equations in this book take this idea and run with it, all the way to the foundations of the connected world you live in. They are about information itself, about how much of it a thing contains and how much you can pour down a wire or fling through the air. And the quantity at their very centre, the thing they measure, carries a name you have just spent a whole chapter coming to understand. They call it, too, entropy.

Chapter Twelve

Shannon-Hartley Theorem

The Ceiling

$$C = B\log_2\left(1 + \frac{S}{N}\right)$$

Right now, a voice can travel from one side of the planet to the other and arrive clear enough to recognise a laugh. A film can pour out of a machine in a distant country and play on the screen in your hand without a stutter. Every second, an unfathomable torrent of messages, voices, pictures, money, love letters, is being chopped into streams of ones and zeroes and flung down copper wires, along threads of glass, and through the open air as invisible waves.

It feels limitless. It is not. There is a hard ceiling

on how much information you can push through any channel, a wall that no invention, no genius, no amount of money can ever break through. It does not matter how clever your engineers are or how advanced your technology becomes. The limit is fixed by the laws of mathematics, and it was found, complete and exact, by one man in one paper in 1948.

You have even heard the sound of someone hitting that ceiling. If you are old enough, you remember the shriek and hiss of a dial-up modem connecting, that strange robotic song. What you were listening to was a machine pushing information down an ordinary telephone line as fast as the universe would allow, and then stopping, because it had run into a wall that Claude Shannon had drawn, on paper, decades before the modem was built. This chapter is about that wall, and it is the most important idea you have never heard of, the equation that quietly governs the entire connected world.

The Question No One Could Answer

Before Shannon, communication was a craft of trial and error. Engineers built telegraphs and telephones and radios, and they made them better by tinkering, by intuition, by guessing. But nobody could answer the two questions that mattered most.

First: is there a limit? Given a particular channel, a telephone wire of a certain quality, say, is there a maximum amount of information it can carry, or can you always squeeze more through with a cleverer trick? Nobody knew.

And second, deeper and stranger: can you ever send a message through a noisy channel *without errors*? Every real channel is noisy. There is always hiss, static, interference, the random jostling of the world corrupting your signal. Everyone simply assumed that noise meant mistakes, and that the only way to cut down mistakes was to slow down, to speak more carefully, to repeat yourself. Send fast and accept errors, or send slow and accept tedium. That was thought to be the iron trade-off of all communication.

Shannon answered both questions at once, and his second answer was so surprising that it took the world years to believe him.

Reading the Equation

Here is the ceiling, the maximum rate at which information can be sent through a channel:

$$C = B\log_2\left(1 + \frac{S}{N}\right)$$

Let us read it piece by piece. C is the capacity, the absolute maximum speed, measured in bits per second, the number of ones and zeroes you can reliably push through every second. That is the ceiling itself.

B is the bandwidth, the width of the range of frequencies your channel can carry. Think of it as the *width of the pipe*. A wider pipe, more bandwidth, lets more through, and it does so in direct proportion: double the bandwidth, double the ceiling. This is why a fibre-optic cable, with its enormous bandwidth, can carry so much more than an old copper telephone wire.

S/N is the signal-to-noise ratio: how strong your signal is compared to the background hiss of noise. S is the power of your signal, N the power of the noise, and what matters is their ratio. A clean, loud signal on a quiet line has a high ratio. A faint signal buried in static has a low one.

And notice that the signal-to-noise ratio sits inside a $\log_2$, a logarithm. This is the crucial and slightly cruel detail. Because of that logarithm, making your signal louder helps only feebly. To add a fixed amount to your capacity, you have to *double* the signal-to-noise ratio, and then double it again, and again. Brute force, just shouting louder, runs into sharply diminishing returns. Widening the pipe pays off far better than shouting. Every engineer who has ever tried to wring more speed out of a connection has felt the weight of that logarithm.

The Proof in the Noise

Why this formula? Here is the intuition, and it is genuinely illuminating, though I will be honest in a moment about how far it goes.

Imagine sending information by setting a voltage on a wire, choosing from a set of distinct levels. With no noise at all, you could use as many levels as you liked, a thousand, a million, infinitely fine gradations, and each choice would carry a huge amount of information. Noiseless, a single wire could carry infinity.

But there is always noise, and noise blurs each level into a fuzzy band of uncertainty. If two levels are closer together than the blur, the receiver cannot tell them apart, and the information is lost. So the real question becomes: how many *distinguishable* levels can you fit, given the blur? Your signal has a certain size, and the noise has a certain size, and roughly speaking the number of levels you can reliably tell apart is set by how the signal compares to the noise. Work it through and the number of distinguishable levels comes out close to the square root of $\left(1 + \frac{S}{N}\right)$.

The information carried by choosing one level out of that many is the logarithm of the number of levels, which gives about $\frac{1}{2}\log_2\left(1 + \frac{S}{N}\right)$ bits for each symbol you send. And the bandwidth B sets how many symbols you can send each second, about $2B$ of them.

Multiply the bits per symbol by the symbols per second:

$$C = 2B \times \tfrac{1}{2}\log_2\left(1 + \frac{S}{N}\right) = B\log_2\left(1 + \frac{S}{N}\right)$$

and the formula falls out.

Now the honesty this book insists on. That argument is a *heuristic*. It gives the right answer, and it shows you why the pieces are where they are, but it is not Shannon's actual proof, and it should never be dressed up as one. Shannon's real proof is deeper, stranger, and far more beautiful, and it leads to a conclusion the counting-levels picture completely hides. So let us go and meet the real result, because it is the part that changed the world.

The Thing That Should Have Been Impossible

Here is what Shannon truly proved, and why it stunned everyone.

He showed that as long as you send information *below* the capacity C, you can communicate with *as few errors as you like*. Not fewer errors. Essentially *zero* errors. Error-free communication, over a noisy, hissing, corrupting channel, as reliable as you please, right up to the ceiling. And the instant you try to push *above* C, reliable communication becomes impossible, and errors flood in without limit. There is a sharp cliff edge at the capacity, and on the safe side of it, perfection is available.

This broke the old iron trade-off in half. Everyone had believed that beating down errors meant slowing to a crawl. Shannon proved that was false. The trick is not to slow down or shout louder. The trick is to be *clever about how you encode the message,* to bundle it up with just the right pattern of extra checking digits, so that even after the noise has chewed on it, the receiver can work out exactly what was sent. He did not say how to build such codes, only that they must exist, lurking, waiting to be found, all the way up to the ceiling.

That single promise launched a seventy-year treasure hunt for those codes, and we found them. It is the reason a scratched CD still plays perfectly, the noise of the scratch erased by error-correcting code. It is why a space probe at the edge of the solar system, its whisper of a signal nearly drowned in cosmic static, can still send back flawless photographs. It is why your phone can hold a clean call in a noisy tunnel, why a torn and smudged QR code still scans, why a video arrives unbroken over a shaky connection. Every one of those everyday miracles is Shannon's promise, kept. He proved the prize existed, and engineers spent two generations claiming it.

Be the Judge

Let us put a number on the ceiling, using the very wall the dial-up modem hit.

An ordinary telephone line carries a band of frequencies about 3,000 cycles per second wide; that is its bandwidth, B. And a decent phone line has a signal-to-noise ratio of about 1,000, meaning the signal is around a thousand times stronger than the hiss. Feed these into Shannon's equation:

$$C = 3000 \times \log_2(1+1000) = 3000 \times \log_2(1001)$$

Now, $\log_2(1001)$ is just asking "two to what power gives about a thousand?", and since two to the tenth is 1024, the answer is very nearly 10. So:

$$C \approx 3000 \times 10 = 30{,}000 \text{ bits per second.}$$

About thirty thousand bits a second. And there it is. That is almost exactly where dial-up modems stalled, screeching away, unable to go faster no matter how the manufacturers tried, because the telephone line itself, by the iron law of this equation, simply could not carry more. The modem makers were not failing. They had driven a hand-built machine right up against a ceiling that Shannon had calculated, on paper, before the transistor was even invented. You can do that sum yourself and find the wall an entire industry spent years pressed against.

The World It Built

Every link in the connected world lives in the shadow of this one inequality. When engineers design a new generation of mobile phones, lay a transatlantic fibre cable, or aim a dish at a satellite, they are always asking the same question: how close can we get to Shannon's ceiling? The whole history of communication technology since 1948 can be read as a long, patient climb toward that fixed line, never crossing it, forever approaching it. The newest codes, with names like turbo codes and LDPC, now hug the ceiling so closely that there is almost no room left to improve. We have very nearly reached the limit Shannon drew, and we know we can go no further, because he proved there is nowhere further to go. There is a strange comfort in that. The pipe has a known size, written in an equation, and the engineers have all but filled it.

Is It Really True?

It is, in the strongest sense available to us, because it is not merely tested but *proven*. Like Pythagoras at the start of this book, the Shannon limit is a mathematical theorem. It does not depend on any experiment that might one day surprise us; it follows by airtight logic from the definition of information itself. Seventy-five years of the most determined engineering in history have climbed toward it and never once breached it, exactly as the proof guarantees. There is no cleverer modem, no future technology, no alien civilisation, that can send information faster than C through a given channel. The ceiling is as solid as arithmetic.

The Name on the Door

The two names in the title, Shannon and Hartley, tell their own small story of how knowledge is built.

Ralph Hartley, working at the same Bell Laboratories two decades earlier, in 1928, had taken the first real steps, suggesting that information could be measured by the logarithm of the number of distinguishable messages. Harry Nyquist, also at Bell Labs, had worked out how fast you could signal down a line. They laid the foundations. But it was Claude Shannon who, in 1948, took their scattered beginnings and built the entire cathedral, adding the one thing they had not mastered, the effect of noise, and proving the astonishing theorems about error-free communication that neither had dreamed of. His paper, *A Mathematical Theory of Communication*, simply created a science where none had existed, almost complete on its first appearance, and the digital age has been living inside it ever since.

And Shannon himself was one of the most delightful minds of the century, a man who carried his genius lightly. He rode a unicycle through the halls of Bell Labs, often while juggling. He built machines for the pure joy of it: a mechanical mouse named Theseus that could learn its way through a maze, one of the earliest glimmers of machine learning; a contraption whose only function, when you switched it on, was to extend a mechanical hand and switch itself off again. He was as much a playful tinkerer as a towering theorist, and he seems to have founded the information age almost as a side project, then wandered off to think about juggling and chess. The man who drew the ceiling over all of human communication did it, you sense, mostly because the problem was beautiful.

So Listen

You are living inside this equation every waking second. The clarity of a voice from overseas, the smoothness of a film streaming to your screen, the fact that your messages arrive whole and not as garbled nonsense, all of it is engineering pressed up against Shannon's ceiling and held there by codes that turn his promise into hardware. The next time a call comes through crystal clear from the other side of the Earth, you might spare a thought for the quiet truth behind it: that there was a limit, that we found it, and that we built our entire connected civilisation right up to its edge.

But there is something we have quietly taken for granted this whole chapter, a word we have used again and again without ever pinning it down. *Information.* We have measured how fast you can send it, how much a channel can carry, where the ceiling lies. We never once said what it actually *is*. How do you measure an amount of information? What makes one message carry more of it than another? How much information is in a word, a coin toss, a face, a genome? That is the final question of this book, and Shannon answered it too, in the very same paper, with one more equation. It is the deepest of them all, it measures the stuff that the modern world is made of, and at its heart sits a quantity with a strangely familiar name. We met it one chapter ago. He called it entropy.

Chapter Thirteen

Shannon's Information Entropy

The Measure of Surprise

$$H = -\sum_i p_i \log_2 p_i$$

We have come a long way, you and I. We began with a right-angled triangle scratched in the dirt, and we have travelled through falling apples and orbiting moons, through the strange circle hidden in five numbers, through the wing that holds up an aircraft and the light that pours from the Sun, down into the blur of the atom and the wall around what can be known, out to the bending of space and time, and into the one-way river that carries us all from birth toward the end. Thirteen equations. The architecture beneath the visible world.

There is one question left, and it is the one the modern age turns on. All through the last chapter we spoke of *information*, how fast you can send it, how much a wire can carry, and we never once stopped to ask what it actually is. It is the most ordinary word in the world now. We swim in information, we are drowning in it, we carry libraries of it in our pockets. But how do you measure it? What makes one message carry more of it than another? How much information is in a single word, a coin toss, a human face, a strand of DNA?

For all of human history, nobody could answer. Information seemed far too vague, too human, too tangled up with meaning, to ever put a number on. And then, in the same astonishing 1948 paper, Claude Shannon answered it, in a single equation, and in doing so he laid the foundation stone of the entire digital world. It is a fitting place to end, because this last equation is, in a quiet way, about all the others. It is about knowledge itself.

Information Is Surprise

Shannon's great move was to throw away meaning entirely.

He realised that to measure information, you must ignore what a message *means* and look only at how *surprising* it is. Information, he saw, is the resolution of uncertainty. It is the amount of surprise a message delivers, the degree to which it tells you something you did not already know.

Picture it. If a friend in the desert texts you "the sun rose this morning," they have sent you no information at all, because you were already completely certain it would. The message resolves no uncertainty. But if they text "it snowed here last night," that carries a great deal, because it was deeply unexpected, and it sharply changes what you know. A message you could have predicted carries little. A message that astonishes you carries a lot. The more unlikely the thing being reported, the more information its report contains.

This is the key that unlocks everything: *information is*

surprise, and surprise is just unlikeliness. A near-certain outcome, when it happens, tells you almost nothing. A rare outcome, when it happens, tells you a great deal. Information and probability are two sides of one coin, and now we can turn that insight into a number.

The Proof in a Logarithm

How much information, exactly, is in an event of probability p? Let us derive it, because the answer is forced upon us by one simple, reasonable demand.

The demand is this: information should *add up*. If you learn two unrelated things, the total information you have gained should be the information from the first plus the information from the second. Learn the result of a coin toss in Sydney and, separately, a coin toss in London, and you have gained exactly twice the information of one toss. Surely that must be true. Information should add.

But here is the snag. When two events are independent, their probabilities do not add, they *multiply*. The chance of two separate coins both landing heads is one half times one half, a quarter, not one half plus one half. So we are caught in a bind. We need a way of measuring information that *adds* when the underlying probabilities *multiply*.

There is exactly one mathematical tool in all the world that does this, that turns multiplication into addition. The logarithm. It is the logarithm's whole defining magic: the log of two things multiplied together equals the log of the first plus the log of the second. So information *must* be built from a logarithm of the probability. There is no choice in the matter; the demand that information add leaves only this one possibility.

So the information carried by an event of probability p is the logarithm of p, with a minus sign in front to make it come out positive, since rarer events, with smaller p, must carry *more* information:

$$\text{information} = -\log_2 p$$

We use logarithm base 2, and that choice gives the unit its name. The information in a single toss of a fair coin, one of two equally likely outcomes, comes out to exactly one. We call that unit a *bit*, the answer to one perfect yes-or-no question, the atom of all information. Every bit on every device you own is one of these, one resolved either-or.

And the entropy, H, the quantity this whole chapter is named for, is simply the *average* information per message from a source, found by weighing the information of each possible outcome by how often it occurs:

$$H = -\sum_i p_i \log_2 p_i$$

That sweeping symbol just says "add up, over every possible outcome, the probability of that outcome times its information." It is the average surprise per symbol. It is the true, irreducible information content of a source, measured in bits. It is, at last, the number we were looking for.

Be the Judge

Let us compute it, with the simplest source there is: a coin.

Take a fair coin, heads or tails equally likely, each with probability one half. Its entropy is:

$$H = -\left(\tfrac{1}{2}\log_2 \tfrac{1}{2} + \tfrac{1}{2}\log_2 \tfrac{1}{2}\right)$$

Since $\log_2 \frac{1}{2}$ is just -1, this becomes:

$$-\left(\frac{1}{2}\times(-1) + \frac{1}{2}\times(-1)\right) = 1$$

Exactly one bit per toss. Each flip of a fair coin delivers precisely one bit of information, the most a single yes-or-no thing can carry. That is the maximum.

Now take a *bent* coin, one weighted to land heads nine times out of ten. Surely it carries less, because you can already guess it will probably be heads. Let us check. With $p = 0.9$ for heads and 0.1 for tails:

$$H = -\left(0.9\log_2 0.9 + 0.1\log_2 0.1\right)$$

$$\approx -(0.9 \times (-0.15) + 0.1 \times (-3.32)) \approx 0.47 \text{ bits.}$$

Less than half a bit. Just as our intuition promised: a predictable coin carries less information, because most of its tosses tell you what you already suspected. And push it to the extreme, a two-headed coin that *always* lands heads, with $p = 1$. Its entropy is zero. A coin that can only do one thing tells you nothing at all when it does it. No surprise, no information, zero bits.

There is the whole idea, computed in your own hand. Maximum information lives in maximum uncertainty, the fair coin, the perfect fifty-fifty, the most unpredictable source. Any structure, any bias, any predictability *lowers* the entropy, because predictable things are less surprising. Information is surprise, and surprise is exactly what the equation measures.

Why Your Files Fit

This is not abstract. It is the reason every device you own works the way it does, because Shannon's entropy is the hard floor of *compression.*

The entropy of a source tells you the absolute minimum number of bits you need, on average, to record its messages. You can be clever and get close to that floor. You can never, ever go below it. And the trick to getting close is exactly the trick the bent coin showed us: spend few bits on the likely, common things, and save your long descriptions for the rare ones.

Consider English text. If all twenty-six letters were equally likely, each would carry nearly five bits. But they are nowhere near equally likely. The letter "e" is everywhere; "q" is rare and almost always followed by "u"; a "t" at the start of a word is very often followed by "h." English is riddled with predictability, with structure, with redundancy. Shannon himself measured it and found that the true information content of English is only about one bit per letter, not five. Which means English text is around three-quarters predictable padding. And that is precisely why it compresses. Every time you zip a file, every time a photo is saved as a JPEG or a song as an MP3, a machine is hunting down the predictable parts, the low-surprise redundancy, and squeezing them out, pressing the data down toward its entropy, its true information content, and no further. The reason your files fit on your phone is that the world is more predictable than it looks, and Shannon gave us the equation that measures exactly how much.

The Equation We Have Met Before

Now look hard at Shannon's equation, and then think back one chapter, to Boltzmann standing among the steam engines, counting the arrangements of atoms.

$$H = -\sum_i p_i \log_2 p_i \qquad\qquad S = k \log W$$

They are the same equation.

Shannon's measure of information and Boltzmann's

measure of disorder, born sixty years apart, one in the study of telephone lines and the other in the study of heat, are, in their mathematical bones, identical. The entropy of a hissing communication channel and the entropy of a box of gas are not cousins or analogies. They are one idea, wearing two coats. The disorder of the universe is missing information. The information in a message is removed disorder. Entropy, it turns out, was always a measure of *what you do not know*, whether you were talking about the jostling of unseen atoms or the unread symbols in a message yet to arrive.

The story goes that Shannon did not even know what to name his new quantity, and asked the great mathematician John von Neumann. Call it entropy, von Neumann is said to have replied, for two reasons: first, because the same formula already exists in physics, and second, because "nobody really knows what entropy is, so in a debate you will always have the advantage." It is a wonderful story, and like several in this book, it may be a little too good to be wholly true. What is certain is that Shannon chose the name deliberately, because his equation had the very same form as Boltzmann's. He was also, to his great credit, cautious, and said for years that he was not sure the two entropies had any deep physical connection at all. They might, he thought, just happen to share a shape.

He was being too modest. The connection turned out to be real and physical, and proving it laid a ghost that had haunted physics for seventy years. Remember Maxwell, back in Chapter Six, and a mischievous puzzle he posed: imagine a tiny demon guarding a door between two halves of a gas, letting fast molecules through one way and slow ones the other, sorting hot from cold for free, building order out of chaos, seeming to break the second law of thermodynamics itself. For generations no one could say exactly why the demon must fail. The answer, when it finally came, was pure Shannon: the demon has to *measure* each molecule, to gather *information* about it, and information is not free. To keep working, the demon must eventually erase its records to make room for new ones, and erasing information, it was proved, *always* costs entropy, dumped as heat into the world. The demon's books balance after all, paid in the hard currency of information. Information is not some airy abstraction floating above the physical world. It is physical. It has a cost, measured in heat and entropy. The bits in your phone and the warmth of your coffee are denominated in the same coin.

Information at the Bottom of Everything

Sit with how strange and large this is. We set out, thirteen chapters ago, to understand the physical world, the triangles and forces and fields and stars. And we have arrived, at the very end, at the discovery that information sits among the deepest things there are, as fundamental as energy, as real as matter, woven into the laws of heat and the fate of the universe.

Some of the boldest physicists of our age now suspect that information may be the most fundamental thing of all, that the universe, at its very bottom, is something closer to a vast computation than a vast machine, that *it from bit,* as one of them put it, that everything physical may arise from answers to yes-or-no questions. When we discovered that black holes, those pits of pure gravity from Chapter Ten, possess entropy, it turned out their entropy counts the information swallowed inside them, written, astonishingly, on their surface. The threads of this book, gravity and heat and the quantum and information, are tangled together far more tightly than anyone building telephones in 1948 could have guessed. We do not yet know where it leads. But it is becoming hard to escape the feeling that information is not something we invented to describe the world. It may be what the world is made of.

Is It Really True?

Like Pythagoras and like the Shannon limit before it, this is a theorem, proved with the certainty of mathematics, not merely tested against the world. Given Shannon's few reasonable demands, the only that information add, that certainty carry none, the equation follows with no escape. And its reach is total. Every act of compression, every error-corrected message, every encrypted secret, every line of code, every triumph of the machines that now learn from oceans of data, rests on this measure of information. It is the bedrock of the entire digital civilisation, the most thoroughly relied-upon equation of the modern age, running silently inside every device on Earth, billions of times a second, right now.

The Name on the Door

The name belongs, almost wholly and very rightly, to Claude Shannon, who in one paper conjured an entire science out of nothing and handed humanity the keys to the information age. He saw, where no one had seen before, that meaning could be set aside and information itself could be counted, and he was correct in nearly every particular on the first attempt, a feat that still astonishes the people who study what he did.

But you will notice one more name standing quietly behind the door, the same name that closed the previous chapter: Ludwig Boltzmann. For the deepest equation of the information age turned out to be, letter for letter, the equation that the tragic

Austrian had found among his atoms and his steam, the equation carved on his tombstone, the one he did not live to see vindicated. The man who explained the arrow of time and the man who measured information never met; they worked lifetimes and disciplines apart. They wrote the same equation. And that, perhaps more than anything, is the secret this whole book has been circling. These thirteen equations are not thirteen separate inventions. They are glimpses, from thirteen different windows, of a single connected structure underneath everything, the architecture that does not care which door you came in by, because it was one building all along.

So Look Around You

We are at the end now, so let me say plainly the thing this book was always about.

You were told, somewhere along the way, perhaps gently and perhaps not, that this was not for you. That mathematics was a locked room, that equations were the private language of a clever few, that you could use the world the experts built but never hope to understand it. There is an old line, often hung on Einstein's name though he never said it, about how everybody is a genius, but if you judge a fish by its ability to climb a tree it will live its whole life believing it is stupid. I opened this book with it, because it names the quiet lie so many people carry: that they are the fish, and the equations are the tree.

But look at what you have just done. You have derived $E = mc^2$ from a box and a flash of light. You have weighed the speed of light out of two numbers measured with wires and magnets. You have divided an apple's fall by sixty squared and landed on the Moon. You have proved the arrow of time with a handful of coins, and computed the information in a bent one. You did not take these on faith. You checked them. You were never the fish. You simply had not been shown the room.

That is the whole of it. These equations are not trusted because clever people signed their names to them; you have seen, again and again, how tangled and human and disputed those names really are. They are trusted because they have been checked, relentlessly, by everyone who doubted them, and because reality keeps voting for them. They have been tested to destruction and have not broken. That is the only authority they have, and it is the only authority worth anything: not the name on the door, but the fact that, when you go and look, the world agrees.

So look around you. The right angle in the corner of the room is Pythagoras, holding the wall straight. The phone in your hand finds you on a map by trusting that gravity bends time, and fills its screen by squeezing data down to its entropy, and holds its signal up against Shannon's ceiling. The warmth on your skin is mass becoming light in a furnace ninety-three million miles away. The very fact that you are solid, that you can hold this book and not fall through the floor, is a wave's refusal to be crushed into too small a space. It was all here the whole time, holding up your entire world, invisible only because no one ever turned and showed it to you.

Now someone has. The architecture was never hidden. It was just unread, and it was written, all of it, in a language you were always able to learn. Go and look again at the ordinary world. You will never quite see it as ordinary again.

Afterword

What Thirteen Could Not Hold

Thirteen is a choice, not an inventory.

This book set out to trace a particular path through the history of physics and mathematics, from the right angle that holds up a wall to the equation that measures a thought. That path is real and it is connected, each chapter growing from the one before it, and I believe it is one of the most remarkable stories our species has ever stumbled into. But it is not the only path, and I want to be honest with you about what it left behind.

There are equations missing from these pages that have shaped the world just as completely as the ones in them.

The Fourier transform, for one. It is the mathematics that says any shape, no matter how jagged or complicated, can be broken apart into simple waves of different frequencies and then rebuilt again perfectly. It is why an MRI machine can reconstruct a picture of the inside of your body from signals that look like noise, why your phone can compress a photograph without losing the detail, why the entire field of signal processing exists. It did not make the thirteen because I had already told its story sideways, through Shannon. But it deserves its own book.

The Navier-Stokes equations describe the flow of fluids, water, air, blood, the atmosphere itself. Every weather forecast you have ever heard, every aircraft wing we designed in chapter five, every simulation of a hurricane, is downstream from them. They are also, quietly, one of the deepest unsolved problems in mathematics. We can use them. We cannot fully prove they behave well. The Clay Mathematics Institute has offered a million dollars to anyone who can. The prize is unclaimed.

The Dirac equation extended Schrödinger's work into the language of special relativity, and in doing so it predicted, before anyone had looked for it, that for every particle of matter there must exist a mirror-image particle of antimatter, identical in every way except carrying the opposite charge. Paul Dirac found antimatter on paper in 1928. Carl Anderson found it in the laboratory in 1932. It is, by most accounts, the purest example in history of mathematics predicting something real before anyone went looking for it.

And the Standard Model Lagrangian, the single most ungainly equation in this book's list of omissions, summarises in one expression everything we currently know about the fundamental particles and the forces between them. It is a triumph, and a mess. It works to extraordinary precision. It requires around nineteen free parameters that have to be measured rather than derived, numbers we plug in because the universe chose them and we do not yet know why. It also has two enormous gaps. It says nothing about gravity. And it says nothing about the dark matter that appears to make up most of the mass of the universe. The most complete description of fundamental physics we have is already, visibly, incomplete.

These are not footnotes. They are the frontier.

And the frontier has shape. Having spent this long with these equations, I can tell you where the walls are, because the walls are the most interesting thing.

Quantum mechanics and general relativity are each fantastically successful, and they cannot both be right at the same time. We have known this since roughly 1930 and we have not fixed it. The theory of quantum gravity, the one that would contain them both, remains unbuilt. String theory and loop quantum gravity are the two most serious attempts. Neither has yet made a prediction that any experiment has confirmed. The problem is not that nobody is trying. It is that the place where the contradiction bites, at scales of energy and smallness so extreme that no instrument we can build comes near them, is effectively unreachable. We are trying to write a theory of the place where our two best maps disagree, without being able to visit the territory.

The arrow of time is another open wound. This book traced it to the low entropy of the Big Bang, and that is the right answer as far as it goes. But it only pushes the question back. Why did the universe begin in such an improbably ordered state? The honest answer is that nobody knows. There are ideas, some beautiful and some very strange, but nothing settled. The deepest fact about your experience of time, the fact that it flows, that you remember the past and not the future, that you grow older and never younger, rests on an initial condition of the universe that remains unexplained.

And then there is dark energy. In 1998, two teams of astronomers measuring distant supernovae found that the expansion of the universe is not slowing down, as gravity would suggest, but accelerating. Something is pushing the universe apart, faster and faster, and we have no idea what it is. We have given it a name, which is not the same as understanding it. It accounts for roughly sixty-eight percent of the total energy of the universe. The thing that makes up most of what exists is, in every meaningful sense, unknown.

I tell you this not to discourage you but because the open questions are, to my mind, the best thing about science. A closed book is a dead subject. Every one of those gaps is an invitation. The person who closes the gap between quantum mechanics and gravity will have done something no one has managed in a hundred years of trying. That person might be reading these pages.

I should say something about why this book exists, and it is not because I am a physicist.

I am not. I have spent my working life in engineering, in the unglamorous and practical business of making networks run and systems talk to each other. I came to the equations in this book late, and from the outside, drawn in by curiosity rather than training. And what I found, when I finally looked properly, was not the forbidding wall of symbols I had been led to expect. It was a story. A long, human, contested, sometimes tragic, often beautiful story about people trying to read the language the universe is written in.

What changed, for me, was not gaining expertise. It was gaining the habit of checking. Of asking not just "what does it say" but "why should I believe it, and how would I know if it were wrong?" That habit, more than any single equation, is what these pages were trying to pass on.

I grew up being handed answers by authority. Most of us did. Some of those answers were good ones. Some were wrong. The useful thing physics teaches is not a set of facts but a posture: hold your conclusions loosely, demand that they earn their place, and let reality have the final vote every time. It is a habit that turns out to be useful well outside physics.

If this book did one thing, I hope it was to show that the equations are not the private property of the people whose names are on them. They belong to anyone willing to check the working. The working is checkable. I showed you the working, and I meant it.

If you want to keep going, the doors are wide open.

Start with Carlo Rovelli's *Seven Brief Lessons on Physics*. It is short, beautiful, and deeper than its size suggests. If the chapter on time stayed with you, read *The Order of Time* next.

For quantum mechanics, Sean Carroll's *Something Deeply Hidden* is one of the clearest ways into the measurement problem, even if you do not end up agreeing with his interpretation. For relativity and modern physics more broadly, Brian Greene's *The Fabric of the Cosmos* remains one of the great works of popular science.

Brian Cox and Jeff Forshaw's *Why Does $E=mc^2$?* is a careful and generous companion to the relativity chapter, and Richard Feynman's *The Character of Physical Law* is still one of the finest short introductions to what physics is really trying to do.

For mathematics itself, read Paul Lockhart's *A Mathematician's Lament*. It says, better than almost anything else I know, why the way many of us were taught mathematics missed the point so badly.

These books will take you a long way past where these thirteen chapters stop.

One last thing, and then I am done.

Somewhere along the way, the story of science became the story of geniuses: singular, towering, unreachable figures who saw things the rest of us could not. That story is wrong, and this book has tried, chapter by chapter, to chip away at it. The names on the doors are complicated. The ideas have many parents. The discoveries were almost always almost made by someone else at almost the same time, because the questions were ready to be answered and the tools were ready to answer them. No equation in this book fell from the sky. Every one of them was built by human beings, arguing, borrowing, making mistakes, correcting each other, and trying again.

Which means the ideas were always within human reach. They were always things a person could find.

You are a person.

The room is still open.

www.ingramcontent.com/pod-product-compliance
Lightning Source LLC
La Vergne TN
LVHW090307210926
848480LV00001B/160